Environmental Systems and Societies SL

FOR THE IB DIPLOMA

Adrian Palmer

PEAK

Published by:
Peak Study Resources Ltd
1 & 3 Kings Meadow
Oxford OX2 0DP
UK

www.peakib.com

Environmental Systems and Societies SL: Study & Revision Guide for the IB Diploma

ISBN 978-1-913433-36-9

© Adrian Palmer 2016–21

Adrian Palmer has asserted his right under the Copyright, Design and Patents Act 1988 to be identified as the author of this work.

All rights reserved. No part of this publication may be reproduced, stored in a retrieval system, or transmitted in any form or by any means, without the prior permission of the publishers.

**PHOTOCOPYING ANY PAGES FROM THIS PUBLICATION,
EXCEPT UNDER LICENCE, IS PROHIBITED**

Peak Study & Revision Guides for the IB Diploma have been developed independently of the International Baccalaureate Organization (IBO). 'International Baccalaureate' and 'IB' are registered trademarks of the IBO.

Books may be ordered directly from the publisher (see www.peakib.com) and through online or local booksellers. For enquiries regarding titles, availability or retailers, please email books@peakib.com or use the form at www.peakib.com/contact.

Printed and bound in the UK by:
CPI Group (UK) Ltd, Croydon CR0 4YY

www.cpibooks.co.uk

Cover image:
This image of the Earth, taken by the Apollo 17 crew in 1972, is known as the 'Blue Marble'. With the sun behind the astronauts there is no noticeable shadow. The original photo has Antarctica at the top, and as a consequence is often reproduced upside-down to match our concepts of north and south.
Source: NASA

Preface

Welcome to this Environmental Systems and Societies (ESS) Study & Revision Guide. For the third edition I made a number of updates to meet the changes brought in to the 2015+ curriculum, and with the rapidly-evolving awareness of and responses to environmental impacts I have added further sources and examples for this new Peak publication. I sincerely hope that it helps you achieve the best possible grade in your final exams.

I also hope the course inspires you beyond getting a good grade. In making a move towards a more sustainable society, we all have something to contribute. Here in Bangkok I have been working on a low impact eco-guesthouse project. The design incorporates a solar chimney (hot air extraction using the solar thermal), geothermal cool air intake by convection, insulating boards made from recycled waste, and solar power for electricity (see Figure 7.5 on page 174).

I dedicate this book to the memory of Dave Tayler (1964–2017), a passionate conservationist, inspirational environmental educator, and friend.

Thanks to many proofreaders, in particular, Charlie Hall who has screened out many mistakes and superfluous material, and others who have read sections of the work.

Adrian Palmer

Contents

Chapter 1: Foundations of Environmental Systems and Societies 1
    1.1   Environmental Value Systems   1
    1.2   Systems and Models   9
    1.3   Energy and Equilibria   13
    1.4   Sustainability   21
    1.5   Humans and Pollution   24

Chapter 2: Ecosystems and Ecology 31
    2.1   Species and Populations   31
    2.2   Communities and Ecosystems   41
    2.3   Flows of Energy and Matter   54
    2.4   Biomes, Zonation, and Succession   62
    2.5   Investigating Ecosystems   71

Chapter 3: Biodiversity and Conservation 81
    3.1   Introduction to Biodiversity   81
    3.2   Origins of Biodiversity   84
    3.3   Threats to Biodiversity   87
    3.4   Conservation of Biodiversity   93

Chapter 4: Water and Aquatic Food Production Systems and Societies 105
    4.1   Introduction to Water Systems   105
    4.2   Access to Freshwater   108
    4.3   Aquatic Food Production Systems   112
    4.4   Water Pollution   119

Chapter 5: Soil Systems and Terrestrial Food Production Systems and Societies 128
    5.1   Introduction to the Soil System   128
    5.2   Terrestrial Food Production Systems and Food Choices   130
    5.3   Soil Degradation and Conservation   135

Chapter 6: Atmospheric Systems and Societies 143
    6.1   Structure and Composition of the Atmosphere   143
    6.2   Stratospheric Ozone Depletion   144
    6.3   Photochemical Smog   150
    6.4   Acid Deposition   153

Chapter 7: Climate Change and Energy Production 158
    7.1   Energy Choices and Security   158
    7.2   Climate Change: Causes and Impacts   164
    7.3   Climate Change: Mitigation and Adaptation   171

Chapter 8: Human Systems and Resource Use	**180**
8.1 Human Population Dynamics	180
8.2 Resource Use in Society	186
8.3 Solid Domestic Waste	191
8.4 Human Population Carrying Capacity	193

Answers to practice questions **200**

References **232**

About this book

Firstly, the book is designed to help you self-test before exams; to do this well you should attempt written answers—even if you are just bullet pointing or planning essay questions. The practice questions are designed to test knowledge required by the curriculum, and underlying concepts that are useful background knowledge. Using my experience as an ESS examiner and teacher of the course, I have tried to incorporate the 'big questions', required skills, and applications where possible. Answers to questions are provided at the end of the text, but it is worth remembering that there is more than one possible answer. I recommend that you use this alongside the IBO course guide, including referring directly to the big questions, and that you consult the most recent past paper questions.

Secondly, whilst writing the text I have been mindful of common issues with the amount of knowledge that is presented in the course. It is important to stress that this is most clearly laid out in the course guide, including understandings, skills, and applications. I am often asked by students about what details to revise for case studies. There is no simple answer, but you need to support general statements with specific details where possible. With required case studies, the more details you can recall the better you are able to answer longer answers and essay questions. Case studies are presented here, with enough detail for you to use where you don't have a better example. But if you already have good examples from your class or course book, then you should choose carefully as to which you want to revise. It is often most useful to revise the ones which you have direct experience of if possible.

Thirdly and lastly, to try and help you get the most from the text, I have addressed each of the required curriculum points sequentially. During my time on the curriculum review team, however, it was clear to us that there are many ways to work through the course, and there is no single correct way. The course is interconnected, and it is helpful to recap related sections together when revising. I have included the following features to help guide you:

	Exam tips and study notes	A combination of suggestions for how to approach your revision of specific topics, guidance on what you need to know for exams and what can gain extra marks, and tips to guard against common errors or misconceptions.
	Key concept/information	Recurring and significant ideas or information that are particularly important to understanding sections of the course.
	Key term	Selected words and phrases that you should be familiar with.
	Practice question	Questions to help you self-test your understanding. These appear after each topic with answers at the end of the book.
	Case study/worked example	Helpful illustrations to explain a concept or problem type.
	Critical thinking	Sections that explore key questions to promote critical thinking around the topic.
	Cross-reference	Links to connected topics in the guide.
	Timeline	Sequences that help to visualise and contextualise important events.
	More information	Pointers to resources outside this guide that will boost your understanding of the topic.

ENVIRONMENTAL SYSTEMS AND SOCIETIES SL

List of practice questions

Environmental Value Systems	6
Systems and Models	12
Energy and Equilibria	18
Sustainability/Humans and Pollution	28
Species and Populations	38
Communities and Ecosystems 1	44
Communities and Ecosystems 2	50
Flows of Energy and Matter	59
Biomes, Zonation, and Succession	68
Investigating Ecosystems 1	74
Investigating Ecosystems 2	80
Biodiversity	82
Biodiversity and Conservation	99
Water and Aquatic Food Production Systems and Societies	117
Water Pollution	125
Soil Systems and Terrestrial Food Production Systems and Societies	140
Structure and Composition of the Atmosphere/Stratospheric Ozone Depletion	148
Photochemical Smog	152
Acid Deposition	156
Climate Change and Energy Production	163
Mitigation and Adaptation	175
Human Population Dynamics	181
Resource Use in Society	188
Solid domestic waste/Human population carrying capacity	196

Chapter 1: Foundations of Environmental Systems and Societies

1.1 Environmental Value Systems

1.1.1 What Is Meant by an Environmental Value System?

An environmental value system (EVS) is the world view, the perspectives or paradigms that shape the ways individuals and societies may think about and approach environmental issues. As such, an environmental value system involves the interrelation of values, attitudes, premises, and arguments that underpin the fundamental beliefs of societies and individuals within it.

Using a systems approach, we can view a model of human world views by inputs that influence it, and outputs that come from it—see Figure 1.1. This model is a simplification of human psychology, societies and behaviour, but it is possible to see common patterns emerging in societies that support its application.

Environmental value systems apply to the whole of the course. It is important to understand the concepts in this unit, and to apply them to case study examples of human decision making and behaviour.

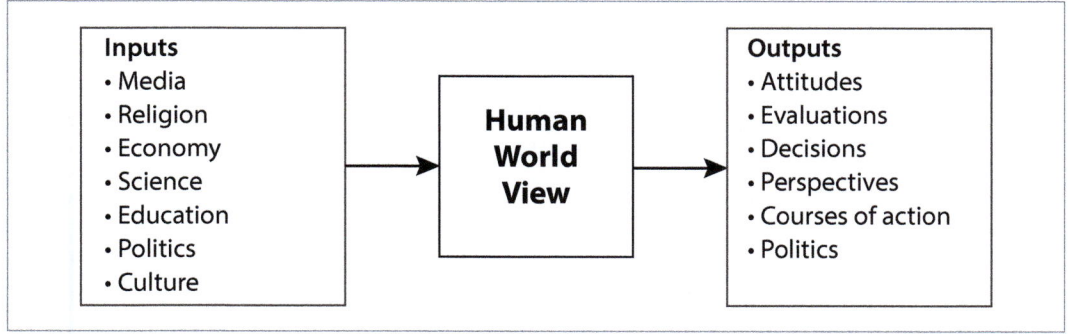

Figure 1.1: **A model of environmental value systems**

1.1.2 Range of Environmental Value Systems

People's attitudes, decisions, perspectives, and actions reveal where they are on this scale for a given situation. In reality, people rarely fit neatly into simple classifications.

The technocentric approach, with a faith that the 'technological fix' can sort out all environmental issues, is often the traditional approach in societies. Within the ecocentric approach, there are a variety of radical, political, philosophical, and spiritual perspectives such as deep ecology. Anthropocentric approaches are typically focused on the human

Stereotypes

Models need to be used with caution when looking at real people. In your extended-response answers, avoid making stereotypical and overly simplistic statements about people's viewpoints.

management capabilities and needs, through existing political structures, governments and organisations. A range of EVSs can be classified and ordered into a scale as shown in Figure 1.2.

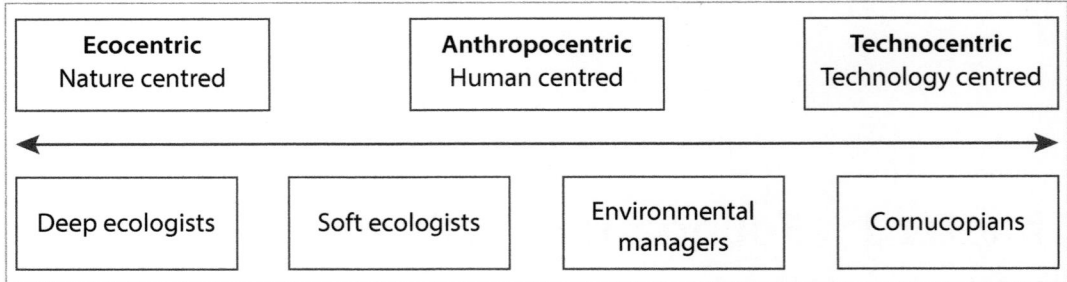

Figure 1.2: **A scale of environmental philosophies**
Source: adapted from O'Riordan, 1981

Read through the summary for each of the three main headings and four world view subheadings shown in Table 1.1. Then consider the application of this model to different environmental issues from the course in terms of solutions.

Environmental Philosophy	Approach to Environmental Issues
Deep ecologists	
Deep ecology is a complex environmental philosophy created by Arne Naess. Adherents believe that a spiritual revolution is needed to fix environmental problems. Nature is at the centre; there should be equal rights for species.	Desire is seen as the driving factor in the whole process of environmental degradation; spiritual impoverishment leads to this desire.
Self-reliance soft ecologists	
They believe that resource management should be self-sufficient and that ecological understanding is a principle for all aspects of living.	Permaculture (permanent agricultural design systems) demonstrates how to integrate diverse farming and resource sharing through local trading systems. They focus on the benefits of small scale economy, in contrast to large scale economies.
Environmental managers	
They work to promote change within the existing social and political structures. They believe that current economic growth can be sustained if environmental issues are managed by legal means or political agreement.	This is the approach of those working within the mainstream of society, e.g., governments and the United Nations (UN). The kind of policy is promoted by NGOs, such as Friends of the Earth. Campaign material is designed for political influence to solve environmental problems.
Cornucopians	
Their view is that environmental issues are not really 'problems', as humans have always found a way out of difficulties in the past. New resources and technologies will solve any environmental problems as they are encountered. There is no need for radical agendas, or socio-economic or political reform.	They are often associated with the idea of frontier economics, to capture then squander the seemingly boundless resources found by pioneers, e.g. the early European settlers in the US Prairies or nomadic slash and burn agriculturalists. It can also be associated with technical scientific approaches to managing the environment and with free market-led development.

Table 1.1: **Summary of some environmental philosophies**

1. FOUNDATIONS OF ENVIRONMENTAL SYSTEMS AND SOCIETIES

1.1.3 Environmental Value Systems and Societies

A society is a group of individuals, connected by geography, historical timeframe, and value systems, for example, the pre-Hispanic society of Peru or Neolithic people of the Western Mediterranean. Political beliefs are often used to characterise and group societies as well, for example, the democratic capitalist societies of Western Europe. Connecting EVSs to named societies is not simple as it is hard to make generalisations that extend to the whole society, particularly in the more multicultural societies of large cities. Learned examples such as those in Table 1.2 will help you describe contrasting EVSs in more depth. If possible, choose examples you have personal familiarity with.

> **Societies**
> Can be defined as groups of individuals connected by common culture, development, location, religion, and value systems.

Buddhist	Judaeo-Christian
Four Noble Truths of Buddhist belief: 1. Suffering exists. 2. Suffering arises from attachment to desires. 3. Suffering ceases when attachment to desire ceases. 4. Freedom from suffering is possible by practising the Eightfold Path. Buddhist monks show a reverence for nature and demonstrate it by living in balance with nature.	Historically Judaeo-Christian beliefs are founded in the Bible and may focus on dominating nature as God gave it to us. *Have many children, so that your descendants will live all over the earth and bring it under their control. I am putting you in charge of the fish, the birds and all wild animals.* Genesis, 1.26 Alternatively, environmentally conscious Christians argue that we have a responsibility to provide better stewardship of the planet as a result of this.
First Nation Americans (Sioux)	**European pioneers (US Prairies)**
The ecocentric approach and lifestyle of the First Nation Americans contrasts with the pioneers who thought this society to be primitive. Most of the many different tribal groups were hunter-gatherers. Many of the Plains Indians, such as the Sioux, were nomadic, and the concept of land ownership was alien to them. From such groups comes the image of the 'noble savage', much loved by environmentalists.	Europeans settlers brought with them fundamental ideas of land ownership that descended from enclosed land in Europe. There was a land grab of the Indian territories; settlers now 'owned' the land. The settlers' attitude to resources was one of frontier economics (see Table 1.1). In the 1890s the pioneers decimated the bison populations and destroyed the ancient way of life practised by the Sioux.
Communist	**Capitalist**
Centrally managed and planned economies could have improved environmental approaches by large scale power generation and state-controlled industry. However, the records of communist governments on environmental management are poor, and they often operated an 'out of sight, out of mind' approach, e.g., the Soviet Union's dumping of nuclear waste into the Arctic Ocean 1951–1991. As there was no free press, citizens were comparatively unaware of environmental issues.	Capitalist democracies have often held governments to account for their actions, with variable degrees of success. Due to a free press and education, awareness of environmental issues grew. However, the approach follows a familiar pattern. First, denial of an issue, then combating, reducing, and delaying legislation. The final approach to environmental issues is to comply with legislation, and operate in the new market for profit, e.g., the response of companies to the issue of stratospheric ozone depletion.

Table 1.2: **Some examples of environmental value systems in different societies**

1.1.4 Application of the Scale

Read the following quote from Al Gore and consider where it fits into the above classification.

> *Right now, it is difficult to imagine that we could cut global emissions of the pollutants that cause global warming by 70 to 80 percent, that the roadways could soon be filled with hybrid electric and plug-in hybrid vehicles, that the greenest buildings could be generating power that they are actually selling back to the utilities. It is perhaps hard now to fully see the potential of hydrogen power, to imagine a smart super-electric grid, to envision new biofuels running our vehicles or having access to the most energy efficient machines and appliances as well as the newest nanotechnologies and manufacturing techniques—all of this would be new to us. But examples of nearly all this technology really do exist today or will exist in the very near future. New market technologies are as difficult for us to see today as the Internet was for workers in the 1980s.*

Gore, 1992, p.xxii

Clearly, these are technocentric solutions being proposed by Gore, but this is not the same as saying that Gore is technocentric.

Evaluate the implications of two contrasting environmental value systems

Choose two EVSs and consider how they may respond to environmental issues you have studied in different ways, how their approaches to managing environmental issues may relate to the value system. Consider this worked example, what typical approach from each philosophy may be seen in the case of climate change?

Ecocentric	Deep Ecologist	The problem must be resolved by reconnecting at a spiritual level to the biosphere in order to reduce consumption.
	Soft Ecologist	Through lifestyle change and sustainable agriculture, carbon footprints can be reduced.
Anthropocentric	Environmental Manager	Through international regulation and government taxation, industry and society will reduce carbon footprints.
Technocentric	Cornucopian	There is no need for international government action; clean technology will arrive on the free market. If the climate warms too much, we can cool it down by geoengineering.

Table 1.3: **EVSs and different solutions to climate change**

Justify, using examples and evidence, how historical influences have shaped the development of the modern environmental movement

Environmental movements have developed all over the world in response to significant historical events. The movements have been triggered by environmental disasters, technological developments, media reports, and significant publications. They are also marked by the response of societies in terms of international agreements (conventions) and legislation (protocols). The timeline shown in Table 1.4 summarises some key events and general trends.

1. FOUNDATIONS OF ENVIRONMENTAL SYSTEMS AND SOCIETIES

You don't need to memorise all the events in the timeline, but you should be able to outline key historical influences on the environmental movement, citing at least three examples in depth. Local examples are acceptable providing they are justifiably significant and balanced with some global examples. It is useful to review the general trend in environment and development by looking at the summary for each period.

Period	Accidents and incidents	Protests, publications, and organisations
1940–59 After the Second World War, awareness is raised of inadequate food production and depletion of resources. Agriculture intensifies.	1956: Minamata deaths from mercury pollution in food chain 1952: Great smog in London kills as many as 12,000 from air pollution	1948: IUCN founded 1958: UN Law of the Sea started
1960–79 Awareness is raised on local air and water pollution issues as by-products of consumption. A broad protest movement begins. The need for environmental stewardship is promoted.	1960: Population: 3 billion; private consumption: 4.8 trillion US dollars 1968: Apollo 8 pictures of 'earthrise' 1969: Cuyahoga river (USA) catches fire due to pollution 1978: Love Canal 1979: Three Mile Island accident raises awareness of risks of nuclear power	1961: WWF founded 1962: Rachel Carson publishes *Silent Spring* (Carson, 1963) 1974: CITES started 1977: Greenpeace 'Save the Whale' campaign 1979: World Climate Conference raises awareness of climate change
1980–99 Awareness is raised of global environmental pollution issues (ozone, acid deposition, anthropogenic climate change). Radical protests start. The need for environmental stewardship becomes established.	1984: Bhopal Disaster: 3,000–4,000 die due to an explosion at pesticide factory in India 1986: Chernobyl disaster 1991: 1 million tonnes of crude oil dumped into Persian Gulf at the end of Gulf War	1980: World Conservation Strategy 1980: Friends of the Earth begins confrontational protest 1997: Seven days taken to evict 'Swampy', a road protester dug underground in the UK
2000–20 Concern broadens on global environmental change, including climate, biodiversity, genetic engineering, deforestation, water management, migration, new and spreading diseases, and globalisation. Many protest groups become mainstream. Corporate environmentalism becomes common.	2000: Population: 6 billion; private consumption: 20 trillion US dollars 2005: Hurricane Katrina hits US Gulf Coast 2007: Great Pacific Ocean Garbage Patch discovered (plastic pollution with a diameter of 2,700km) 2011: Fukushima Daiichi nuclear disaster following a tsunami in Japan 2015: SE Asian haze causes air pollution crisis in 8 countries 2019: SARS coronavirus type 2 infects people in China, leading to a global pandemic that kills over 5 million people in the first two years 2020: Consequence of abrupt climate change now regularly observed in events worldwide, including record wildfires occurring in the US, Australia and the Arctic	2001: US rejects Kyoto Protocol 2006: Al Gore's *An Inconvenient Truth*, directed by Davis Guggenheim. 2009: G-20 London summit 35,000 protestors march for 'Jobs, Justice, and Climate' 2012: Rio +20 Conference or Earth Summit, leads to the Sustainable Development Goals 2015: Paris, UN Climate Change Conference 2015: UN Sustainable Development Summit, the publication of the report *Transforming our world: the 2030 Agenda for Sustainable Development* and finalisation of the Sustainable Development Goals 2020: Widespread growth of activism for climate justice and action, including school climate a strikes (Fridays for the Future) and direct action to shut down transport, commercial areas and oil refineries (Extinction Rebellion)

Table 1.4: **Over 80 years—a timeline of key environmental events and societies' responses**

ENVIRONMENTAL SYSTEMS AND SOCIETIES SL

Discuss the view that the environment can have its own intrinsic value

Environments and their biospheres are valued by societies in different ways. Economists, philosophers, and environmentalists trying to value nature have developed the idea of intrinsic value. These values are unrelated to human beings altogether; it is the value that the environment and life forms have in their own right. If you need clarification consider your answer to this question: without humans, do the environments and living species of the planet have value? If your answer is yes, you are attributing an intrinsic value. If people didn't exist, the intrinsic value would remain unchanged. The argument can be extended to entire environments and ecosystems (link this to deep ecology), including non-living components such as rocks.

Intrinsic values are fuzzy, in other words, hard to define. This means that economists struggle to award these financial values. Individuals give different valuations of intrinsic value depending upon cultural, ethical, spiritual, and philosophical perspectives.

Practice questions: Environmental Value Systems

1. *State* what is meant by an environmental value system.

   ....

   ....

   ....

   ....

Command terms

When answering any IB question, do what the command term says. I have formatted command terms such as *state* so they stand out, and they are defined in the course guide. Tips are also written in boxes in this guide.

2. *Evaluate* the application of systems modelling to humans in Figure 1.1.

   ....

   ....

   ....

   ....

3. *Outline* the range of environmental philosophies by completing the following table:

Environmental Philosophy	Outline of Key Ideas, People, or Actions
Deep Ecologists	
Soft Ecologists	
Environmental Managers	
Cornucopians	

1. FOUNDATIONS OF ENVIRONMENTAL SYSTEMS AND SOCIETIES

4. ***Construct*** your own timeline of environmental events and use it to outline key historical influences on the development of the modern environmental movement.

   ....
   ....
   ....
   ....
   ....
   ....

5. ***Evaluate*** the implications of two contrasting environmental value systems in the context of:
   - Human population growth (8.1.2)
   - Global food supply (5.2.1 and 4.3)
   - Energy supply choices (7.1)
   - Water resource management (4.2)
   - Conservation of biodiversity (3.4)
   - Management of a named conservation area (3.4.8)
   - Ecological footprints (1.4.8 and 8.4.4)
   - Love Canal (8.3.5)
   - Eutrophication (4.4.6)
   - Photochemical smogs (6.3)
   - Depletion of stratospheric ozone (6.2)
   - Acid deposition (6.4)
   - Sustainable development (1.4.3)

   ....
   ....
   ....
   ....
   ....
   ....

6. Discuss the view that the environment can have its own intrinsic value. Use examples to illustrate your answers.

   ....
   ....
   ....
   ....
   ....

ENVIRONMENTAL SYSTEMS AND SOCIETIES SL

7. Look at the ten quotations. **Construct** an environmental scale using Figure 1.2 and rank their positions along the scale.

....

....

....

....

....

....

> **Quotations for questions 7 and 8**
>
> **A.** Necessity is the mother of invention.
>
> Anon
>
> **B.** Humankind has not woven the web of life. We are but one thread within it. Whatever we do to the web, we do to ourselves. All things are bound together. All things connect.
>
> Chief Seattle, 1855
>
> **C.** The end of the human race will be that it will eventually die of civilization.
>
> Ralph Waldo Emerson
>
> **D.** There has been a fundamental misconception that conservation means nothing but the husbanding of resources for future generations.
>
> Gifford Pinchot (Callicot, 1993)
>
> **E.** One wonders whether deep ecology's biocentric maxim that all living beings can be equatable with one and another in terms of their 'intrinsic worth' would have any meaning during the long eras of organic evolution before human beings emerged. The entire conceptual framework of deep ecology is entirely a product of human agency—a fact which imparts to the human species a unique status in the natural world.
>
> Murray Bookchin (Bookchin, 1993)
>
> **F.** I would sooner expect a goat to succeed as a gardener than expect humans to become responsible stewards of the Earth.
>
> James Lovelock
>
> **G.** We abuse land because we regard it as a commodity belonging to us. When we see land as a community to which we belong, we may begin to use it with love and respect.
>
> Aldo Leopold (Leopold, 1949)
>
> **H.** There's no scientific proof that global warming even exists. To be honest, it's a bogus consensus dreamed up by Greens because they hate industry. They hate advancement. They hate technology...Greens will lead us back to the stone ages.
>
> Jonathan Hoenig (Fox News, 2006)
>
> **I.** The most serious, insidious danger in the environmental movement is that it may sap the will of advanced communities to face the problems which no doubt lie ahead. Throughout history, hope for the future has been a powerful incentive for constructive change.
>
> Maddox, 1972 (O'Riordan, 1981)
>
> **J.** Back to the Pleistocene!
>
> Earth First! Slogan (Bookchin, 1989)

1. FOUNDATIONS OF ENVIRONMENTAL SYSTEMS AND SOCIETIES

8. Choose three quotes that you agree with and three you disagree with. Justify your personal viewpoint in relation to the quotes.

....

....

....

....

....

....

....

1.2 Systems and Models

1.2.1 The Concept and Characteristics of Systems

A system can be defined as an entity with at least two components (parts) that are linked and interact in some way.

Systems occur in a wide range of settings and scales; systems approaches can be applied to both social and ecological situations.

Systems have emergent properties; these are properties greater than the sum of the systems parts, for example, life can be seen as an emergent property of functioning systems of living organisms.

Systems theory

Systems theory applies to the whole of the course as it is useful in the study of complex environmental issues.

1.2.2 Examples of Systems

In biology there are organ systems, such as the digestive system. The components are the various organs such as the stomach and the small intestine, which work together in an interlinked system. In economics we may identify systems on a variety of scales, such as the functioning of a local market economy or a national banking system, up to the global economy.

In the next unit the concept of the ecosystem is introduced. These naturally occurring systems exist on a variety of scales, from the microscopic world of decomposers in the soil ecosystem to the global ecosystem or biosphere. The biosphere is the portion of the Earth that contains life, from deep in the crustal rocks to high up in the atmosphere. Within the biosphere there are numerous examples of environmental systems, including atmospheric systems and the geochemical cycles such as carbon, nitrogen, and water.

1.2.3 Three Types of System: Open, Closed, and Isolated

Open systems allow energy or matter to enter or leave the boundaries of the system. A clear example would be an animal body. Energy and matter enter as food into the system and both leave the system too.

Closed systems allow energy to enter or leave but matter is contained within the boundaries. The Earth is an imperfect example of this as some matter enters the atmosphere (such as meteorites) or leaves (such as astronauts).

Isolated systems do not allow either matter or energy to enter. This is largely a theoretical concept used in the laws of thermodynamics, with the only possible example being the entire universe.

1.2.4 Transfer and Transformation of Matter and Energy

Transformation of matter refers to a change in state between solid, liquid, or gas. Transfer of matter is a movement of the matter from one place to another. Energy transformations refer to a change in the type of energy, for example, from light to heat. Energy transfers occur when the same kind of energy moves from one location to another, for example, from heat to heat.

1.2.5 Systems Diagrams

Systems diagrams are models of systems and they have a particular format. They use squares or boxes to represent stores (stocks), and the units are generally energy, mass, or volume. They use arrows to represent flows (inputs and outputs), and the units must be the same as for stores but include time—they are a rate.

In the biosphere the biogeochemical cycles approximate to closed systems, for example, the global carbon cycle. This is shown in Figure 1.3 as an example of a systems diagram.

1.2.6 The Concept and Characteristics of Models

Models are simplified representations of the real world. They may help us think clearly about what occurs in the real world and to help us analyse how the real world may respond. However, they are not the real world and so weaknesses and limitations occur in all models as a result of simplification.

1.2.7 Examples of Models

A quantitative systems diagram can be used as a model of systems behaviour—see Figure 1.3 for an example. In a quantitative systems diagram it is possible to see what the quantities of matter or energy are in the stores and flows.

Various types of models are also used to show interactions, more generally, through flow charts. In this guide general models of interaction have ovals instead of boxes—see, for example, the models of feedback (1.3.5) or eutrophication (4.4.6). If you look at these carefully, you will see the arrows represent general interaction and not the flow of energy or matter.

There are a wide variety of other techniques employed to describe models clearly, including graphs and diagrams—see, for example, the demographic transition model (8.1.4), the daisyworld model (1.3.11), and the niche model (2.1.6).

Matter: anything that takes up space and has mass. It is normally made of atoms.

Energy: the ability to do work and affect either the transformation (change) or transfer (movement) of matter. Energy lacks mass but can be found stored in chemical form. Other forms of energy relevant to environmental systems include heat, light, and movement (kinetic).

1. FOUNDATIONS OF ENVIRONMENTAL SYSTEMS AND SOCIETIES

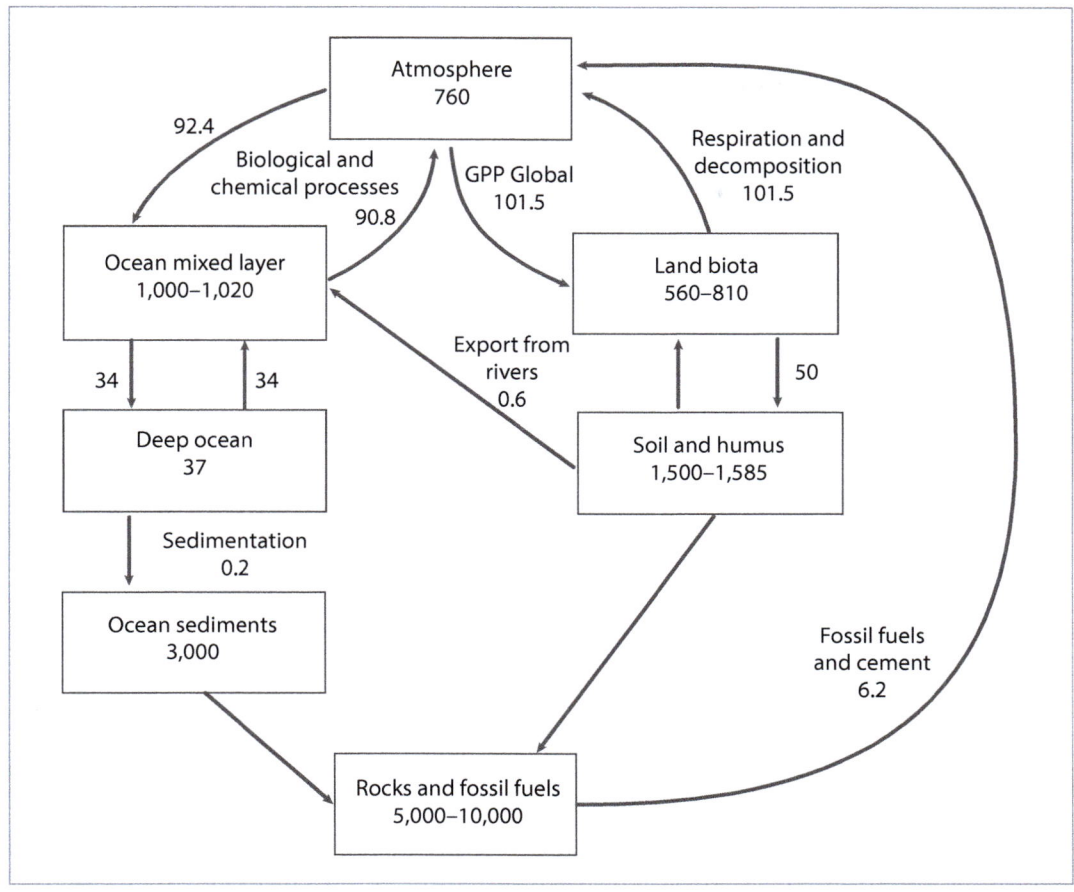

Big numbers

Carbon stores are shown in gigatonnes (Gt) and flow in Gt per year.

1 Gt = 1×10^9 metric tonnes = 1,000,000,000 t
1 Gt = 1×10^{12} kilograms = 1,000,000,000,000 kg

Notes

Ocean mixed layer is the surface water.
GPP is gross primary productivity (2.3.8).
Biota refers to all living things.
Humus is dead organic material in soil.

> **Quantities**
> Quantitative details are shown in Figure 1.3, but you don't need to learn quantities. You should be able to understand and analyse them.

Figure 1.3: **A system diagram of the global carbon cycle**
Source: Adapted from OU, 1986; Wright & Nebel, 2002; Canadell & Dhakal, 2010

1.2.8 Evaluating Models

Making predictions based on models is particularly difficult in environmental systems, or societies, as there are so many factors and feedback interactions. There are also thresholds at which a system's behaviour can change and settle into a new equilibrium (tipping points).

When you are asked to evaluate methods or models, you need to explore both the strengths and the weaknesses of the technique. To help with this, consider the following points:

- What are the limitations of knowledge? (The data may not be known or not available.)
- Is the data a representative sample of the whole system? Is the data reliable?
- Is it an evidence-based model that makes logical sense?
- Are there feedbacks with possible tipping points?
- Are there potential influences on the model that are missing?
- Is the model a useful tool for analysing the situation?

Practice questions: Systems and Models

1. ***Outline*** the concept and character of a system.
   - Give some examples from
   - Environmental sciences
   - Other subject areas
   - Real (everyday) life

   ....
   ....
   ....
   ....
   ....
   ....

2. Give three examples of the following different kinds of systems:
   - Open
   - Closed
   - Isolated

   ....
   ....
   ....

3. ***Define*** the terms 'matter' and 'energy'.

   ....
   ....
   ....

4. ***Define*** and give examples of models. ***State*** how they can be represented.

   ....
   ....
   ....
   ....

1. FOUNDATIONS OF ENVIRONMENTAL SYSTEMS AND SOCIETIES

1.3 Energy and Equilibria

1.3.1 The First Law of Thermodynamics or the Law of Conservation of Energy

This law states that the total energy of the universe or any isolated part of it will be the same before and after matter is moved or transformed.

In other words, energy can neither be created nor destroyed, merely changed from one form to another. For example, during the burning of coal, stored chemical energy is transformed to movement, heat, light and sound. The total amount of all energy transformed by combustion (movement + heat + light + sound), must be equal to the starting amount of chemical energy.

> **Thermodynamic Laws**
> The laws of thermodynamics govern the flow of energy and ability to do work in ecosystems. They define and limit ecosystems' emergent properties.

1.3.2 The Second Law of Thermodynamics or the Law of Entropy

This law states that in an isolated system the total amount of entropy (disorder) will tend to increase.

So, when energy is transformed there is a loss of order or complexity. This is seen practically by loss of heat energy during processes like the combustion of coal above. Chemical energy is more ordered than heat energy.

As energy transformations all involve heat loss, the ultimate extension of the law is to predict that all other forms of energy will eventually transform to heat and cool down—the so called 'heat death' of the universe.

1.3.3 The Laws of Thermodynamics and Environmental Systems

So, in summary of these laws, we can see that the universe is made of matter that is affected by the forces of various energies. These energies move or transform the matter. In doing so, the energies cannot be used up but change from one kind to another. In time, an isolated system will tend to become more disordered. So, how can complex life exist on the planet, seemingly in contradiction of this law?

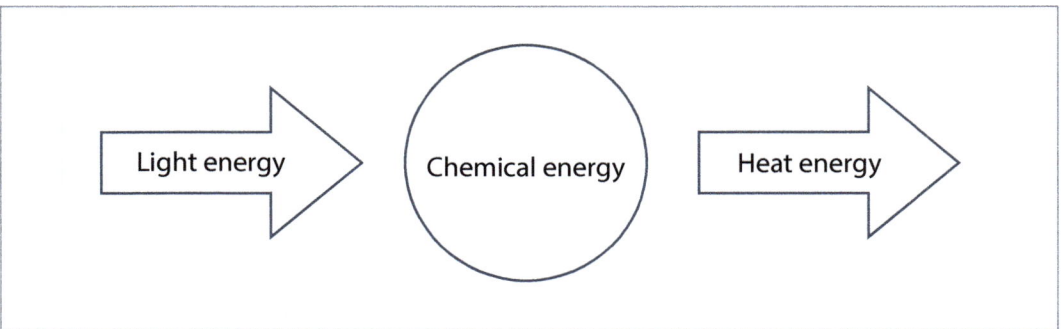

Figure 1.4: **A simple model of energy flow on the Earth**

The Earth's systems are not isolated. They import energy and use this to work against the entropy law. This allows living organisms and ecosystems to build up their order and complexity, due to the constant energy input from the sun. If this input were to fail then the complex living structures of the planet would start to obey the second law, increasing in entropy. The end result would be the breakdown of all ordered chemical structure through death and decay.

All energy stored in chemicals will be finally released as heat energy radiating out to space. The laws apply to all environmental systems, from the carbon cycle to the hydrological cycle, to energy flow through the ecosystem, or changes taking place during succession. Consequently, you are advised to reread this section when looking at systems found in later units and consider how the laws apply to those units.

1.3.4 The Nature of Equilibria

Equilibrium is a condition where forces or quantities are in balance. Therefore, two equal masses on pivoting scales are in equilibrium. This example is a static equilibrium, and there is no change to either side.

Ecosystems are open systems that may be found in equilibrium, such as the numbers of individuals in each trophic level, or the proportions of mass in different compartments of the ecosystem. There may be a flow of matter or energy between components in the ecosystem. For example, nitrogen may move in either direction between the soil store and the atmosphere store, but the quantities are in balance. So, there is equilibrium and the amount in the stores should not change, even though there is movement. This is termed a steady-state equilibrium.

Many environmental systems show resilience to change. This means that, following a disturbance, the system has the capacity to return to the same state of equilibrium. This form of stability is enhanced by negative feedbacks (see below).

1.3.5 Feedback Loops

Feedback mechanisms produce an effect on the equilibrium of a system. Flows in feedbacks are circular cycles, returning to produce an effect on the initial change. Both natural and human systems are influenced by feedback mechanisms, which may involve time lags. There are two kinds of feedback: negative and positive.

1.3.6 Negative Feedback

An initial change in the system leads to a cycle that returns to, and reduces, that change. Negative feedbacks are important in maintaining a stable equilibrium in a system. These feedbacks work in both directions and each arrow shown below can have many steps.

> **Equilibria**
> Equilibria are found as stable states of a system, but systems may have several alternative stable states. Systems pushed past tipping points will move to a new stable state.

> **Apply the general to the specific**
> Use the general diagram shown here, along with the examples described in 1.3.7 to explain how a decrease in x (temperature), may lead to an increase in x in different situations.

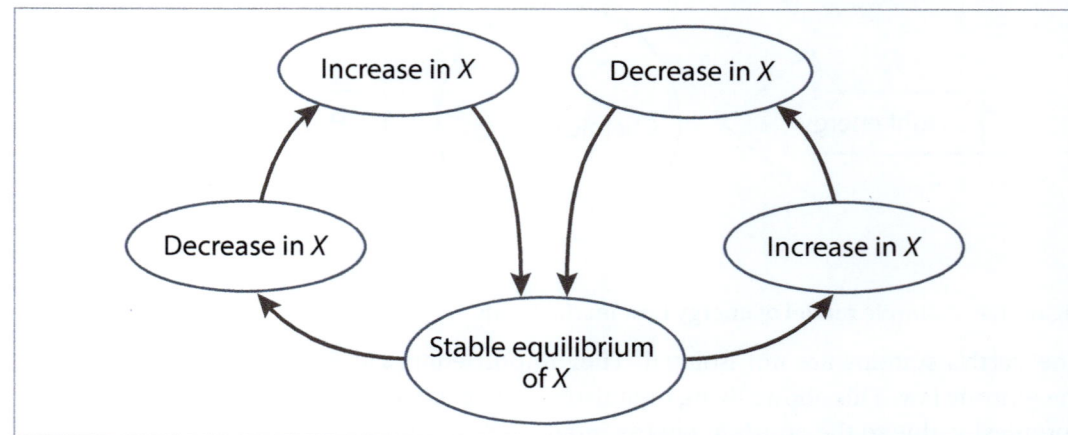

Figure 1.5: **Generalised flow diagram of negative feedback loops**

1. FOUNDATIONS OF ENVIRONMENTAL SYSTEMS AND SOCIETIES

1.3.7 Negative Feedback in Temperature Regulation

Temperature regulation is often maintained by negative feedbacks in natural and artificial systems. The air conditioning thermostat will switch the cooling mechanism on when the temperature rises above a set point, controlling temperature by negative feedback. After the room cools, then the thermostat switches off. Cooling systems occur in biology, for example, the bodies of mammals are controlled in a similar way to thermostats. Blood temperature is monitored by the brain: if it is too cool then warming mechanisms come into action; and, if too hot, the body's cooling system 'switches' on until the temperature is back to equilibrium. As an ecological example, the daisyworld model (1.3.11) shows how a similar process operates in the global ecosystem. In ecological communities negative feedbacks are important in regulating populations interacting through predator-prey cycles (2.1.12).

1.3.8 Positive Feedbacks

Positive feedbacks are produced by an initial change in the system triggering a cycle that returns to, and amplifies, that change. Positive feedbacks produce change in a system; they move away from equilibrium to a new state.

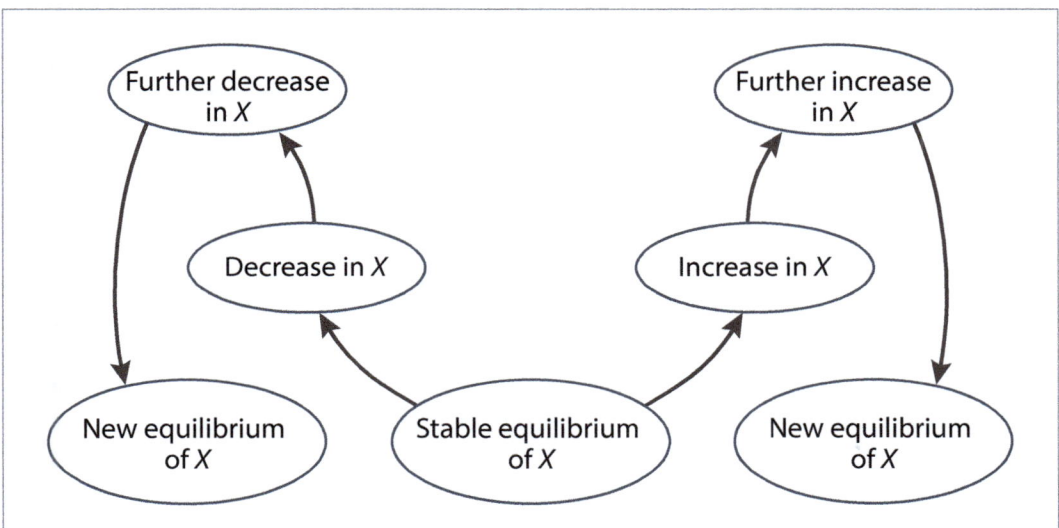

Figure 1.6: **Generalised flow diagram of positive feedback loops**

1.3.9 Positive Feedback in Climate Change and Succession

Positive feedbacks change systems over time, spiralling away from initial equilibrium. Such feedbacks in the global climate system lead to an increase in warming.

For example, the melting of the polar ice caps will lead to a decrease in global albedo, as white ice and snow cover reflects more light. This would lead to an increase in warming as the ground absorbs more heat, leading to a decrease in ice cover.

Positive feedbacks work on an ecosystem that is undergoing succession. Changes in microclimate and soils lead to changes in community structure, diversity, and productivity. At climax, the community remains in balance unless there is disturbance

Albedo: refers to the reflectivity of a surface; black is lowest, and white is highest.

such as tree fall, disease, or human interference. See section 2.4.4 for further detail on succession.

1.3.10 System Stability

Resilience is an aspect of system stability; it refers to the ability of the system to return to its original state. Systems that are more resilient can recover from considerable disturbance, but those that are less resilient may never return to their original state.

Figure 1.7: **Resilience and tipping points**

In Figure 1.7, three possible equilibrium states for a system are modelled as coloured balls. The purple equilibrium has very low stability; it has low resilience as it is close to tipping points, therefore, a small disturbance could cause a large change. The green equilibrium has far greater resilience, unless it is moved beyond the purple tipping point. The orange equilibrium point is the most stable, due to the size of disturbance required to shift it to a new tipping point.

Systems may be locally stable but globally unstable. In Figure 1.7, the purple point has the same local stability as the yellow, but the global stability of the yellow is greater than the purple. Stability is also described by other system properties; they are also said to show resistance or inertia to change, as some systems are disturbed more easily than others. Systems may also be said to show properties of fragility or robustness, depending on internal properties of the system (Table 1.5).

Resistance	How hard it is to disturb the system; low resistance is easily disturbed.
Resilience	How easily the system can return following disturbance.
Robustness and Fragility	Internal properties of the system; robust systems survive in a broad range of conditions and fragile ones in a narrow range.
Local and Global	Local stability means the system can return from a small disturbance; global stability means that it can return from a large disturbance.

Table 1.5: **Generalised flow diagram of negative feedback loops**

Stability can be influenced by system properties such as diversity and the size of stores, both of which can produce significant time lags on system behaviour. As humans can reduce both diversity and the size of the stores, they can impact on the system resilience.

To understand how the size of the store can create time lags, consider this familiar example: the bath tub (Meadows, 2008). This simple system has one input and one output of water: the tap and the drain. If the input is greater than the output, then the bath fills up. But if the bath is filled and the tap is off, the bath will empty when the plug is pulled.

However, the bath will remain filled for some time; in other words, a time lag is present and it is influenced by the size of the stored water in the bath.

Human changes to diversity also impact on system stability. This influence has been debated for a long time, and it is often considered that diverse systems may be more stable. However, there is evidence from mathematical models, and from the occurrence of real systems that are both simple and stable, that this may not be the case. Resilient systems such as the Peruvian rainforest and tundra have both high and low diversity, yet they have both remained as stable systems for a long time, which indicates that diversity is not the only requirement for stability.

1.3.11 The Gaia Theory and Daisyworld

James Lovelock's Gaia theory suggests that feedback mechanisms are important in regulating conditions on a planetary scale. His model of 'Daisyworld' is used to show how populations of white and black daisies could regulate temperature on an otherwise barren planet. White daisies keep cool when it is hot; therefore, they survive and reproduce. Their population grows and so the planet's albedo increases. The planet cools down following a population boom of whites. As it cools down the black daisies gain the advantage as they absorb more heat and keep warm; therefore, they survive and reproduce. As they reduce its albedo, the planet warms up further.

See if you can spot these trends on the graph below. In this world the sun has some random changes in solar energy output, which affects the temperature. Daisies influence albedo; white daisies cool the planet; dark ones warm it. Daisy populations are shown at the top; temperature is shown underneath. Temperature is in arbitrary units.

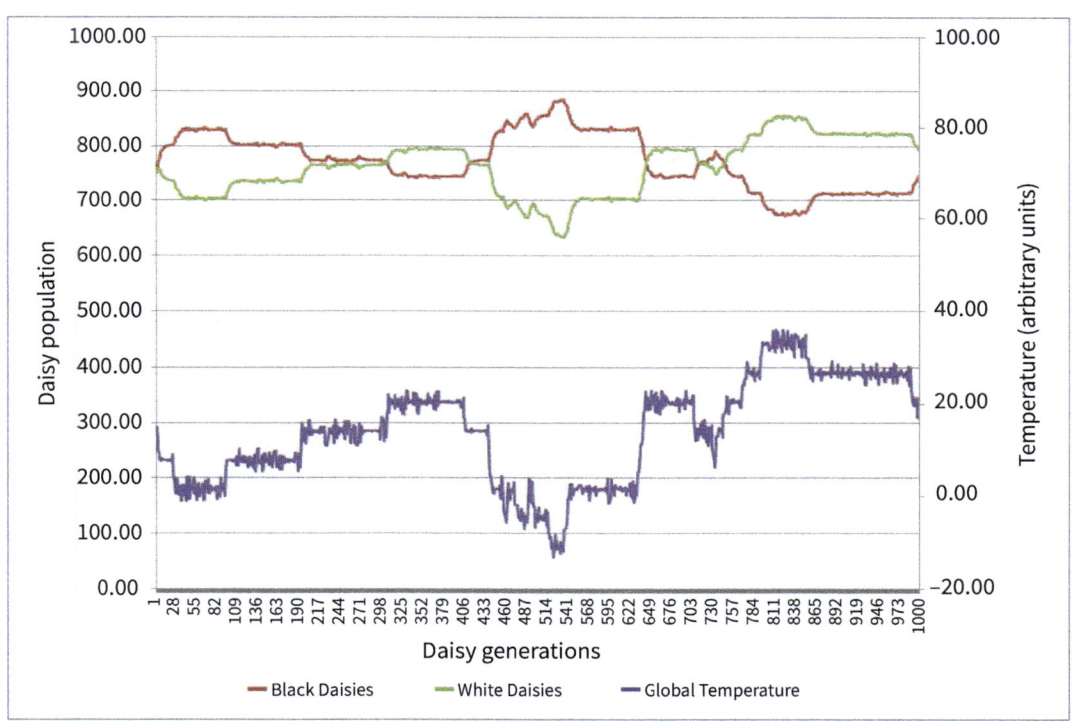

Figure 1.8: **Graph showing temperature regulation, generated by a simulating spreadsheet based on the Daisyworld model**
Source: Based on Lovelock, 1989, using original data

ENVIRONMENTAL SYSTEMS AND SOCIETIES SL

Practice questions: Energy and Equilibria

1. *State* the first law of thermodynamics (energy conservation law).

   ....

   ....

2. *State* the second law of thermodynamics (entropy law).

   ....

   ....

3. *Describe* how the first and second laws of thermodynamics are relevant to environmental systems.

   ....

   ....

   ....

   ....

4. During the process of photosynthesis, a plant uses sunlight to make chemical energy. Heat energy is lost. Assume no other energy changes take place, and then *calculate* the energy value of the sunlight if there are 168 joules of heat and 12 joules of chemical energy produced.

   ....

5. Use the laws of thermodynamics to *explain* why the plant in the previous question grows and increases in complexity. And why the same plant in a sealed box—that neither energy nor matter can enter—would die and decrease in complexity.

   ....

   ....

   ....

   ....

6. *State* which kind of feedback is more likely to lead to stability: positive or negative?

   ....

1. FOUNDATIONS OF ENVIRONMENTAL SYSTEMS AND SOCIETIES

7. **State** if a positive or negative feedback is involved for the following examples:

    (a) Increase in predator population leads to a decrease in prey.

    Decrease in prey population leads to a decrease in predators.

    ....

    (b) Increase in primary production (plant growth) leads to an increase in falling leaves.

    Increase in leaf fall leads to an increase in soil organic matter.

    Increase in soil organic matter leads to an increase in soil nutrients.

    Increase in soil nutrients leads to an increase in primary production.

    ....

    (c) Increase in deforestation.

    Increase in deforestation causes erosion, loss of nutrient, flooding and drought.

    Increase in erosion, loss of nutrients, flooding and drought leads to a decline of crop yields.

    Decline in crop yields leads to an increase in poverty.

    Increase in poverty leads to increased deforestation from clearing land.

    ....

8. **Draw** simple flow models to show an example of one negative and one positive feedback.

9. **State** the symbols used to represent flows and stores in a systems diagram and the difference between flow (process) units and storage units.

    ....

    ....

10. *Calculate* the annual change in atmospheric carbon predicted by the model in Figure 1.3.

   ....

   ....

   ....

   ....

11. *Identify* if there is a steady-state equilibrium shown in Figure 1.3 on page 11 for the interactions between:

    (a) Atmospheric and ocean mixed layer store

    ....

    (b) Atmospheric and land biota store

    ....

    (c) Atmospheric and all other carbon stores

    ....

12. *Evaluate* the systems diagram of the global carbon cycle in Figure 1.3 by completing this table:

What are the limitations of knowledge?	
How reliable or representative is the data?	
Is it an evidence based model that makes logical sense?	
Are there feedbacks, with possible tipping points?	
Are there possible influences on the model that are missing?	
Is the model a useful tool for analysing the situation?	

13. *Draw* a model to show feedback loops in Daisyworld.

1. FOUNDATIONS OF ENVIRONMENTAL SYSTEMS AND SOCIETIES

14. *Evaluate* Daisyworld as a model for predicting temperature change on Earth.

....

....

....

....

15. *Apply* the systems concept on a range of scales, from small scale to global.

....

....

....

1.4 Sustainability

1.4.1 The Lens of Sustainability

> *It is obvious that the real wealth of life aboard our planet is a forwardly-operating, metabolic, and intellectual regenerating system. Quite clearly we have vast amounts of income wealth as Sun radiation and Moon gravity to implement our forward success. Wherefore living only on our energy savings by burning up the fossil fuels which took billions of years to impound from the Sun or living on our capital by burning up our Earth's atoms is lethally ignorant and also utterly irresponsible to our coming generations.*
>
> R. Buckminster Fuller, 1969, p. 94

Any system can be considered through the perspective of sustainability.

The quote above illustrates the principle of, and the need for, sustainability on Earth. Sustainability means that renewable resources must be given time to regenerate after exploitation and any damage to the ecosystem is given time to recover.

Any society, or segment of society, whose growth and development is dependent on reducing essential natural capital is not sustainable. Although this sounds simple, there are different perspectives on this concept that vary from this resource-based definition. Many alternative definitions will be found on the internet, but these are not needed.

The essential requirement of sustainability means living within the natural income without permanently reducing the natural capital. Natural capital are the resources; natural income is the yield we take from these resources.

1.4.2 Natural Capital and Natural Income

Natural resources taken from living or non-living systems for human use can be referred to as natural capital. These resources are naturally occurring goods and services. Goods are physical commodities and services are activities that are used as they are produced.

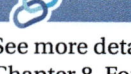
See more details in Chapter 8. For examples of the different kinds of natural capital, see Table 8.3.

Natural capital can produce a natural income in the form of goods or services, such as those listed above. This can be seen as similar to interest on capital savings. Natural income can be calculated as sustainable yield. The goods or services may translate to a direct market place value, for example, timber yield; others may be viewed by society as a free resource, for example, absorbing pollution.

1.4.3 Sustainable Development

This political concept was first outlined in the Brundtland Report, *Our Common Future* (1987). The report recognised and identified the need for more sustainable development and gave one of the most commonly used definitions of the term: 'Sustainable development is development that meets the needs of the present without compromising the ability of future generations to meet their own needs (World Commission on Environment and Development, 1987, p. 43).

Sustainable development has been criticised as the term is contradictory, vague, and flexible and that, 'In implying everything sustainable development arguably ends up meaning nothing' (Adams, 2006, p 3).

> **Sustainable development**
>
> The concept of sustainable development is a key aspect to evaluating management and policy decisions throughout the course.

Economic	Development is economically efficient, and benefits are distributed fairly across generations.
Social and Cultural	Development does not cause conflict and leads to greater control of decision making (empowerment). The range of cultural values within a society should not be endangered by development.
Environment and Ecology	Ecological life support systems; natural capital and biodiversity should not be damaged by development.

Table 1.6: **Aspects or pillars of sustainability**
Source: Reid, 1996; Huckle, 1996

The three pillars model is a useful analytical tool that describes these aspects and can give further structure to the debate and help consider the balance or lack of balance towards sustainable development (Adams, 2006). Try drawing a model with either pillars or overlapping circles for environment, economy, and society. This kind of diagram illustrates different perspectives on the perceived priority of each aspect. You could compare this to other versions such as the original in Adams 2006, or search online for images showing 'three pillars of sustainability'.

1.4.4 Environmental Indicators

Numerous factors are measured and used to indicate environmental change, often combined with general indicators on progress towards sustainability and published by governmental, inter-governmental, and non-governmental organisations. Trends and targets are then used by policy makers around the world to take action on environmental issues at all levels on the local to global scale.

These environmental indicators are diverse and include factors such as:

- Population (for example, crude birth rates)
- Biodiversity (for example, extinction rates)
- Pollution (for example, sulphur dioxide concentrations)
- Climate (for example, incidence of extreme weather).

Many examples of such indicators are published in print and online, for example:
- Sustainable Development Indicators by the UK Government Office for National Statistics (ONS, 2015)
- Vital Signs by the Worldwatch Institute (Worldwatch Instiute, 2015)
- Proposed Global Indicator Framework for the UN Sustainable Development Goals (UN, 2016).

1.4.5 Millennium Ecosystem Assessment (MA)

The UN launched the MA in 2001, providing the largest assessment ever of global ecosystem health, designed to provide a scientific consensus on the state of ecosystems for decision makers in government, business, and civil society.

Many environmental indicators were applied to rate the impact of economic development on ecosystem services. For example:

- Percentage habitat loss between 1950–90 was estimated in terrestrial biomes. The report found that agriculture primarily had removed habitat by over a half of the area in six major biomes.
- The year of maximum catch in fisheries, globally, was found to have been reached between 1965–95, with all fisheries, globally, now in decline.
- Ecosystems services were classified as enhanced, degraded, or mixed. The assessment found that 60% of ecosystem services were degraded.
- Nitrogen flows into rivers and in terrestrial ecosystems were found to have increased; a doubling in flows since the 1960s was recorded.
- Species extinction rates were assessed against the distant past; a 100–1000 fold increase was identified.
- Freshwater extraction for irrigation was estimated with around 15–35% identified as being unsustainable.

The MA aimed to achieve political legitimacy, scientific credibility, and a clear focus on the needs of the users. The assessment concluded that humans have radically altered ecosystems over the last 50 years and that these changes threaten the development goals. Though ecosystem degradation could get worse, with careful management the trends could be reversed. However, solutions to the problems identified will need significant changes in policy (MA, 2005).

1.4.6 Environmental Impact Assessment (EIA)

EIAs are procedures required by the planning processes of most countries. They are produced alongside a development proposal. Though EIA procedure varies from country to country, the steps in an EIA usually include a baseline study of all relevant aspects of the environmental, social, and economic impacts. The purpose of the EIA is to assess the likely impacts before the development and to suggest alterations to mitigate these. The EIA also includes a monitoring programme to take place during and after construction and operations.

1.4.7 Evaluating EIAs

Whilst EIAs provide decision-makers with useful background information, the standards with which they are carried out vary considerably. In some parts of the world the EIA may be carried out by the developer and open to accusations then of bias. In many countries EIAs are not required at all, or regulations on how they are carried out allow developers to find loop holes. Some impacts may fall outside of the scope of the EIA, with some indirect impacts harder to assess. Whatever the findings of an EIA, socio-economic factors are likely to impact on final decisions made, as the EIA process may add considerable expense to development.

1.4.8 Ecological Footprints

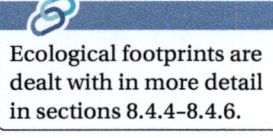

Ecological footprints are dealt with in more detail in sections 8.4.4–8.4.6.

The ecological footprint is the area of land and water in the same vicinity as the population that is required to absorb the population's pollution and waste and supply its resources at a given standard of living. Ecological footprints provide an ecological snapshot of current resource use and natural regeneration. They are useful tools to indicate sustainability, as when they exceed the available land area this indicates a lack of sustainability.

Ecological footprints have allowed estimates of our planetary footprint, in other words, the total human need for land. The estimates in 2010 by the Global Footprint Network show a world average of 2.7 ha/person, although there is 1.8 ha/person actually available. In other words, we currently use approximately 1.5 planets worth of resources in a year. Alternatively, what we use in a year would take one year and six months to regenerate (Global Footprint Network, 2011).

Countries can be compared to this baseline figure: footprints in some countries are over 10 (for example, Luxembourg) and in others under 1 (for example, Malawi). It is apparent from the calculations that some countries are living within their limits while many are in deficit. Gross Domestic Product (GDP) per capita is a measure of the economic wealth of a country divided by the number of people in that country. When nations are divided into groups by GDP the following pattern is shown.

	Population (million)	Footprint (Ha/person)	Examples
High GDP	1,000	6.1	Finland, Belgium, and Australia
Middle GDP	4,300	2.0	Mexico, Uzbekistan, and Bolivia
Low GDP	1,300	1.2	Haiti, The Gambia, and Laos

Table 1.7: **Per capita ecological footprints by GDP**
Source: Adapted from Global Footprint Network

Although, commonly, the footprint is found to be proportional to income status, there are many exceptions of countries with a high income and low footprint.

1.5 Humans and Pollution

1.5.1 Nature of Pollution

An agent means input of energy into the environment, for example, thermal pollution of freshwater or light pollution into a natural habitat.

Pollution is defined as the release into the environment of a substance or an agent by human activities at a rate at which it cannot be rendered harmless by natural processes.

1.5.2 Types of Pollution

Pollutants are produced by human actions and have effects on ecosystems, which may be long term. Note that pollutants may be materials released (matter or substances) or energy released (agent), for example, noise pollution.

Pollution can be **point** or **non-point**, as summarised in Table 1.8.

1. FOUNDATIONS OF ENVIRONMENTAL SYSTEMS AND SOCIETIES

Type	Definition	Management Strategies and Issues
Point source	Pollution released from a single source, for example, sewage effluent released by pipe into a stream, or smoke from a factory chimney.	• Easier to monitor emissions at source • Easier to control emissions at source • Responsibility easily established and managed by law • Localised effects can be managed
Non-point source	Pollution released from diffuse sources, for example, pesticides from farmers' fields or many single sources such as the exhausts of cars in a city.	• Monitoring requires extensive survey techniques • Emissions control requires widespread changes • Responsibility shared amongst many requiring greater effort to enforce change • Effects are spread over a wider area

Table 1.8: **Point and non-point pollution sources**

Pollution can be termed acute when caused by sudden events such as oil spills, or chronic when caused by an ongoing process, or regular release, into the environment—for example, oil released from sea water flushing of oil tankers.

Organic pollutants are from natural sources and usually break down—they are biodegradable. Biodegradable pollution, such as oil or sewage, will be broken down by the action of bacteria or other processes in the environment.

Inorganic pollutants are synthetic chemicals, some of which are also persistent (they do not react and alter form), for example, pesticides and heavy metals such as mercury. Persistent pollutants do not break down easily in the bodies of living organisms or in the environment.

Primary pollutants are effective as soon as they are released into the environment; secondary pollutants are produced through some form of chemical or physical change that alters the primary pollutant.

1.5.3 DDT

DDT (dichloro-diphenyl-trichloroethane) is a powerful insecticide that has been used to control pests in agriculture and for mosquitoes. The key events in its use are summarised in the timeline.

DDT is enough; you don't need to learn the full name.

The rise, fall, and rise again of DDT

1938	DDT rated as the 'magic bullet' of insecticides. • Broad spectrum • Non-toxic to humans and other mammals • Highly persistent—giving long-lasting effect
1940s	Effectively used in the battlefields of the Second World War to control body lice and the disease typhus.
1955	WHO commences a programme to control malaria globally using DDT. Successful eradication achieved in some countries, for example, Taiwan.

ENVIRONMENTAL SYSTEMS AND SOCIETIES SL

1960s	Widespread use of DDT in aerial spraying of crop land, to control pest outbreaks in forests and to control mosquito vectors of malaria in natural wetland ecosystems.
1962	*Silent Spring*, a book by Rachel Carson, raises awareness of the dangers of persistent organic pesticides (POPs). She specifically shows that DDT bioaccumulates and causes thinning of the egg shells in birds of prey such as the peregrine falcon.
1970s & 80s	DDT banned in many developed countries for agricultural usage. WHO switches to malathion in 1976 due to an increasing resistance in mosquitoes. A de facto global ban.
2001	Stockholm Convention on POPs regulates all DDT use. DDT banned internationally for agricultural use but not disease control. The intention is to find alternatives for disease control by 2020.
2006	WHO reverses 30-year policy; DDT now recommended for regular treatment of buildings in areas of high transmission. WHO aims to cut usage globally and reach the Stockholm target of a total phase out by 2020.

1.5.4 Debating a Global Ban on DDT

Currently DDT is banned for agricultural use, but it is used for mosquito control to reduce transmission of diseases. Mosquito-borne diseases such as malaria, dengue, and Zika can be reduced by using DDT. A global ban is debated as part of the Stockholm Convention. The main points are summarised in Table 1.9.

For	Against
Research has linked DDT to premature births, low birth weight, and abnormal mental development of infants.	WHO states DDT is safe if used properly.
Alternative methods of pest control exist; DDT is not the only available pesticide.	Alternatives are not as effective; DDT significantly reduces deaths from malaria when used.
Spraying cannot eradicate the 20 species of Anopheles mosquitoes from native habitats without disastrous impacts on other invertebrates and biodiversity.	Annual deaths from malaria are still over 1 million; 85% are in Africa. The number of malarial cases globally is over 240 million.
The ecological effects are well documented, but the effects of accumulation in human tissue are not fully known.	Previous decisions to ban DDT saw a resurgence of mosquitoes and a rise in deaths from malaria in many countries.

Table 1.9: **Some key points in the debate for banning DDT**

1.5.5 Behaviour of Persistent Pollutants

Persistent pollution may build up in the food chain and the most toxic effects are found in the consumers at the end of the chain. This process of biomagnification (2.2.13) occurred with the pesticide DDT as detailed above; another example is of mercury poisoning in Minamata, Japan. An agri-chemical company released effluent into the bay where it settled onto the sea bed. The first victims were cats which suffered nerve spasms and paralysis, symptoms of neurological damage. Later, people who had eaten the shellfish showed the same symptoms and around 50 people died before the problem was controlled.

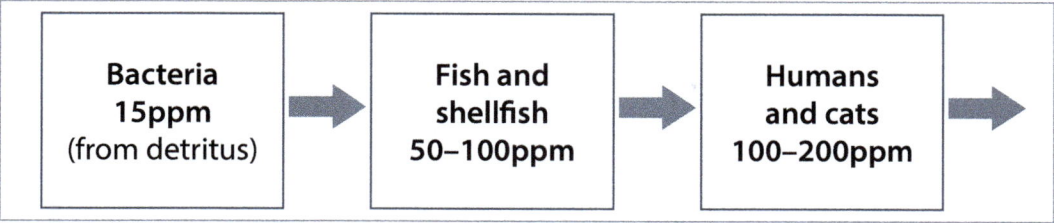

Figure 1.9: **Biomagnification of mercury in the food chain of Minamata Bay, Japan (ppm = parts per million)**

1.5.6 Pollution Management

Management strategies to reduce the negative impacts of pollutants can take different approaches at different levels in relation to the problem. At the initial level, the activity producing the pollution is managed: the ultimate cause. At the next level, the pollution released into the environment is managed: its release (or transfer/transformation). At the final level, the approach is to reduce the negative impacts the pollution is having on the environment, in other words, to manage the effects of the pollution.

The kind of approach taken by societies to manage pollution will depend on the type of pollution, along with many other socio-economic factors. Some possible strategies at three different levels are summarised in the table on the next page.

Level	Against	Management Strategies (examples of some ways in which the impact can be reduced at each level)
Cause	Polluting activity by humans	• Alternative technologies • Alternative life styles • Reduce, reuse, recycle
Release and transfer (transformation for secondary pollutants)	This is the point of emission or the area in the case of non-point. It also includes the pathway or route along which the pollutant may travel.	• Setting standards • Monitoring and imposing standards • Cleaning up emissions
Effects	Long term impact of the pollutant on the ecosystem	• Removing the pollutant from the ecosystem • Replanting or restocking with animals • Habitat restoration

Table 1.10: **Pollution management strategies at three levels**

1.5.7 Applying Systems Thinking to Pollution

It can be useful to apply systems thinking to pollution problems and their solutions. Quantitative systems diagrams can be useful for predictive modelling, such as Figure 1.3 of the carbon cycle. Quantitative approaches are useful for assessing critical loading of pollutants in different ecosystems. Modelling systems through flow diagrams are useful for investigation interaction and feedback, providing a deeper understanding of how the system will respond. See Figure 4.6 for an example of feedbacks during eutrophication.

ENVIRONMENTAL SYSTEMS AND SOCIETIES SL

Practice questions: Sustainability/Humans and Pollution

1. *Explain* how the following quote illustrates the concept of sustainability in terms of natural capital and natural income.

   > *The nation behaves well if it treats the natural resources as assets which it must turn over to the next generation increased, and not impaired, in value.*
   >
   > Theodore Roosevelt

2. *List* some ecosystem services and discuss their value to society.

3. *List* some environmental indicators used in the millennium impact assessment. *Evaluate* the value of these to improving sustainability.

4. *Evaluate* the use of EIAs in attaining sustainability.

5. *Explain* how ecological footprints can be used to show sustainability.

1. FOUNDATIONS OF ENVIRONMENTAL SYSTEMS AND SOCIETIES

6. **Discuss** which aspects of sustainability are covered by these statements:

   Sustainable development is economic development that lasts.

   David Pearce (Pearce, Markandya, & Barbier, 1989)

   ....

   ....

   ....

   Scientific and technological innovations, such as in agriculture, have been able to overcome many pessimistic predictions about resource constraints affecting human welfare. Nevertheless, the present patterns of human activity accentuated by population growth should make even those most optimistic about future scientific progress pause and reconsider the wisdom of ignoring these threats to our planet. Unrestrained resource consumption for energy production and other uses, especially if the developing world strives to achieve living standards based on the same levels of consumption as the developed world, could lead to catastrophic outcomes for the global environment.

   Joint statement made before the Rio summit
   (National Academy of Science and Royal Society of London, 1992)

   ....

   ....

   ....

   A dynamic process which enables all people to realise their potential and improve their quality of life in ways which simultaneously protect and enhance the Earth's life support systems.

   (Forum for the Future, 2011)

   ....

   ....

   ....

7. **Suggest** which if any of these statements improve on the Brundtland report definition given previously. *Explain your answer.*

   ....

   ....

   ....

   ....

   ....

8. *Evaluate* the arguments for and against the use of DDT in agriculture and disease control.

   ....

   ....

   ....

9. *Construct* a systems diagram to show the impact of one example of pollution you have studied.

10. Briefly *describe* pollution management strategies for DDT at each level.

    ....

    ....

    ....

11. *Evaluate* the strategies at each level.

    ....

    ....

    ....

Chapter 2: Ecosystems and Ecology

Ecology is the science of the relationships between living things and the non-living environment. This includes the study of ecosystems and their communities.

Ecosystems are complex systems. They involve the interaction between the living and the non-living components of a defined unit. Ecosystems vary in scale, from the whole world, or ecosphere, to smaller systems like soil systems.

Ecosystems are a functional unit in ecology, units that transform energy and recycle matter. The study of the way that matter and energy are handled by ecosystems is a key concept in environmental systems and societies.

> **Species examples**
> Species names should be *clear* and *detailed*. For example, 'fox' actually refers to over 12 species with very different ecology. Therefore, Cape fox, or red fox, are more specific.

2.1 Species and Populations

2.1.1 Definitions of Key Terms

Term	Definition	Examples
Species	A group of organisms that interbreed and produce fertile offspring. They also share common features.	The tiger, *Panthera tigris*, is considered a single species with six subspecies surviving and three extinct.
Population	A group of organisms of the same species living in the same area at the same time and which are capable of interbreeding.	The number of Indochinese tigers, *Panthera tigris corbetti*, resident in South East Asia has been estimated at around 1,000 individuals.
Habitat	The environment in which a species normally lives.	The habitat of the Indochinese tiger, *Panthera tigris corbetti*, includes a variety of types of forest.
Niche	The set of biotic and abiotic factors that influence a species' distribution. An organism's ecological niche depends not only on where it lives but also on what it does.	The tiger's niche is that of a top carnivore, a major predator on animals like the wild pig, *Sus scrofa*, or other large mammal prey between 20–100kg.

Table 2.1: **Definitions of key terms used in species and populations**

> **Latin names (binomials)**
>
> Although Latin binomial names are not a *requirement*, some examples may **help you gain higher grades**. If you use them in an essay you should underline them; only use capital letters for the genus, as in *Vulpes* for foxes. In typed text, *italics* can be used: *Vulpes vulpes* is the red fox and *V. chalma* is the Cape fox (Genus abbreviated to *V.*).

Factor or component?

Factors should not be confused with the components, although they are related. The components are the parts of the system; the factors are the interactions between these components.

2.1.2 Abiotic and Biotic Distribution Factors

A number of factors have an influence on the existence and, therefore, distribution of a species. These can be divided simply into biotic and abiotic factors. An analysis of these factors is particularly useful to find out about the patterns of distribution of a species, as species are adapted to a specific range of conditions.

2.1.3 Identifying Abiotic Factors

The significance of abiotic factors varies from one ecosystem to another. This table is divided up into three broad categories. You should be able to list the factors for a specific ecosystem you have studied.

Terrestrial	Freshwater	Marine
Light wavelengths are selectively absorbed in forest canopies and in water at depth. This may lead to a change in species such as red algae on the rocks of a shaded woodland stream or deeper in the ocean		
Precipitation varies considerably due to seasons, altitude, and distance from the sea	**Temperature** shows locations, seasonal changes and patterns with depth, with sudden change marked by a thermocline	
Temperature varies due to geographic locations and seasons		
Soil factors vary with local geology, successional stage, and climate. Includes pH, % water content and organic matter	**Turbidity**: cloudiness of the water caused by suspended sediment or phytoplankton	
Wind speed varies according to geographical location, aspect, and shelter from vegetation	**Dissolved oxygen** (mg/l) varies with depth, temperature, and biochemical oxygen demand (BOD)	
Aspect/Slope: the angle and direction of slope influences drainage and microclimate	**Flow rate**: caused by gradient and shape of the river channel	**Salinity**: major changes in intertidal areas, estuaries and where large rivers meet the sea
Elevation: the height above sea level (not altitude, which is height above ground level)	**pH** varies according to local geology or acidification due to pollution	**Wave action** varies from exposed to sheltered shores

Table 2.2: **Summary of abiotic factors in different ecosystems**

2. ECOSYSTEMS AND ECOLOGY

2.1.4 Identifying Biotic Factors

These biotic factors arise as interactions between populations of living organisms, some over long time periods.

Biotic Factor	Definition	Detailed examples
Competition	Populations use resources such as space, light, mates, food, or nutrients are finite (a fixed amount). If they are used, there is less available to others and they may become limiting factors. Competition can be intraspecific (within species) or interspecific (between species).	Duckweed, *Lemna spp.*, and the mosquito fern, *Azolla spp.*, compete for space on the surface of freshwater. In the UK the otter and the introduced North American mink are in competition for food around rivers.
Parasitism	This occurs when species live closely together, but one of the species gains at the other's expense. Parasites may consume body parts or fluids, or capture the host organism's own nutrients from its gut.	Ectoparasites (feeding on the outside of the body) include the leeches and mosquitoes. Endoparasites (feeding on the inside of the body) occur in many ecosystems. In the tropical reef environment each species of sea cucumber is likely to be supporting a pearl fish that enters its body and feeds on its internal organs.
Mutualism	These are relationships between species where both benefit.	Lichens operate as a mini-ecosystem and live in extreme habitats, such as Antarctica. Corals are also good examples, such as the staghorn coral (*Acropora spp.*) and zooanthellae (a kind of algae). Colonies of coral animals filter feed for floating detritus, and the algae carry out photosynthesis.
Herbivory	This is the consumption of autotrophs by a primary consumer. *Note: autotrophs are not always plants but include seaweeds and other algae.*	Limpets (*Patella spp.*) graze on seaweeds (algae) such as the gutweed (*Ulva intestinalis*). There are many other examples such as leaf monkeys or langurs (e.g., *Presbytis spp.*) and browsing the fruit and leaves of fig trees.
Predation	This is the consumption of a primary consumer by a secondary consumer or higher. Eats animals (note—not other organisms). Complex feedbacks lock predator and prey into cycles.	Tigers, *Panthera tigris*, and wild pigs, *Sus scrofa*, live in rainforest, crab-eating macaques (monkeys) and fiddler crabs in the mangroves, Lynx and snowshoe hare in the Canadian tundra.

Table 2.3: **Definitions and examples of biotic factors**

2.1.5 Limiting Factors and Carrying Capacity

Limiting factors are biotic or abiotic factors that may limit the growth of the population, for example, when there is plenty of food, water, and suitable shelter for breeding then populations are free to grow rapidly. If there are predators, parasites, and disease in abundance, or an absence of food, water, and shelter, then populations may not grow easily or go into decline.

You need to be able to apply your knowledge of theory against case studies. Test your understanding of niche against the case study in 3.3.11

There is a natural limit to any population in a defined area of the environment. This limit can only be exceeded on a temporary basis and then the population will go into decline. This limit is termed the environment's carrying capacity, defined as the population that can be sustainably supported in an area. Ultimately, for any species, this depends on the abiotic and biotic interactions of all the species in the environment.

2.1.6 Niche Model

For any species there is a measurable range of a given factor that it can live within. For example, the temperature range of a species has a maximum and minimum tolerance limit, and so for this factor the possible niche of the species can be easily defined as lying within those limits as shown in Figure 2.1.

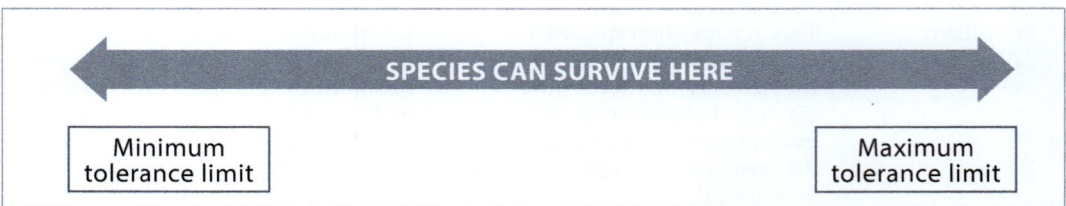

Figure 2.1: **Tolerance limits**

2.1.7 Fundamental Niche

On any species there will be more than one abiotic factor operating. Two factors could be indicated as shown in Figure 2.2. In this model, the centre has optimum conditions (marked by the darker green). At the extremes of tolerance, the organisms will be stressed, which will most likely lower tolerance limits and produce a circular distribution pattern. This model demonstrates the fundamental niche, the set of conditions in which the species could potentially live sustainably, both surviving and reproducing successfully within the abiotic limits.

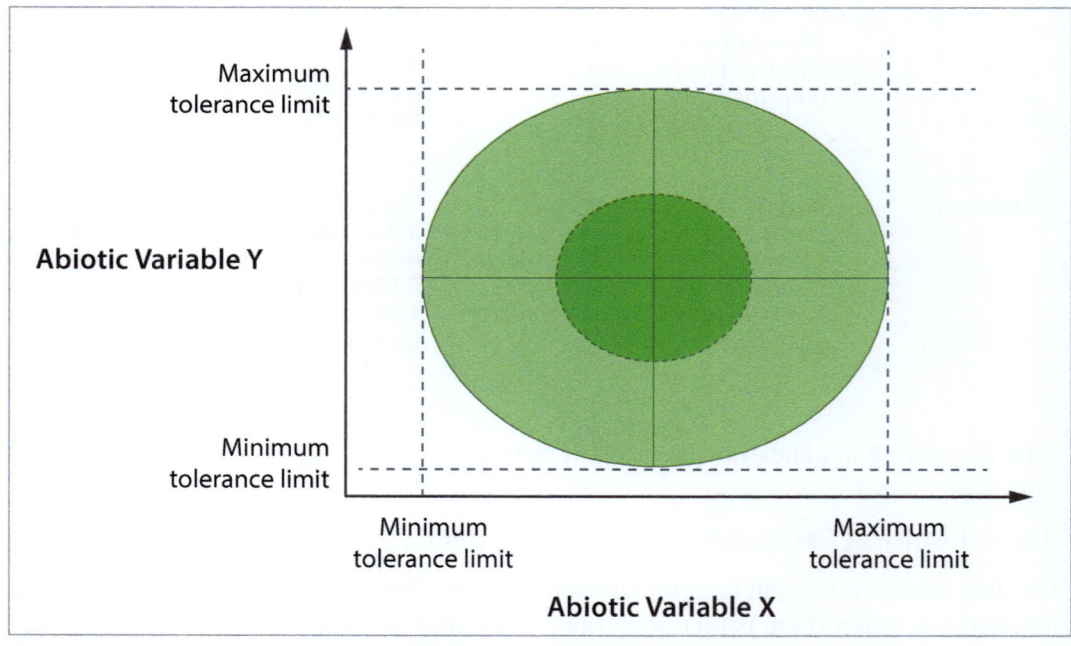

Figure 2.2: **Fundamental niche model showing two abiotic factors**

2.1.8 Realised Niche

Figure 2.3 shows the interactions between two abiotic and two biotic factors. Species will find it harder to survive in the areas overlapping with predators and competition (orange areas). The area in which the species can live sustainably is now within the dark and light blue areas with the bold border. This area is called the realised niche; this is where the species is actually likely to be found in large numbers. Further details of this niche model can be found in Hutchinson, 1957, or Smith, 1995.

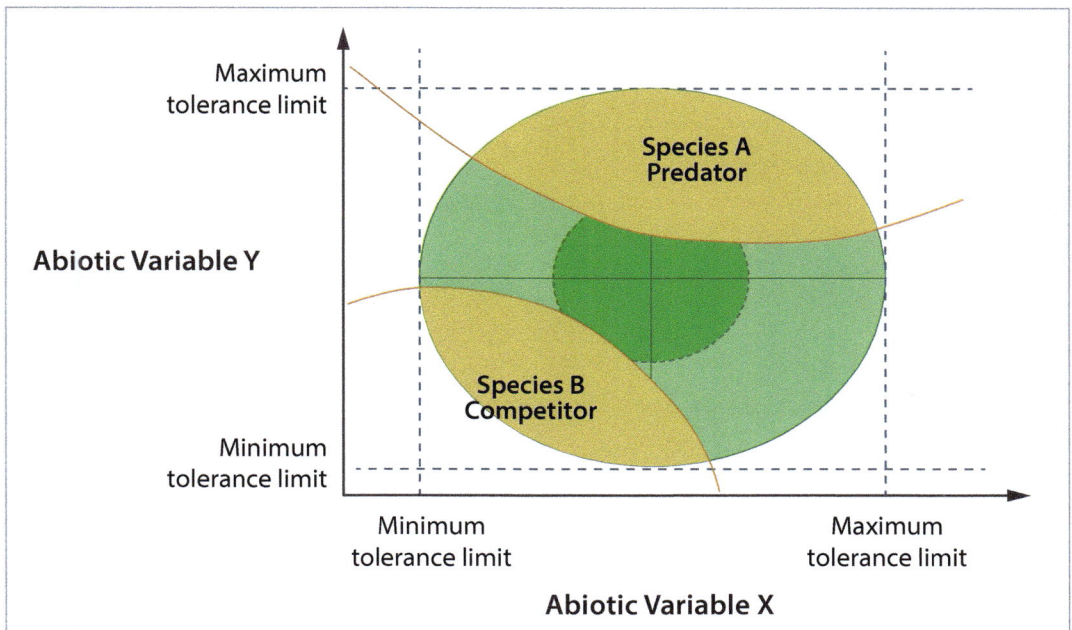

Figure 2.3: **Realised niche model with biotic interactions**

These are simple models that recognise only a few factors. In real ecosystems there may be larger numbers of factors. When evaluating such models, it is important to understand that the interactions occur at the organism level, but the ultimate impact is upon the species population. The total carrying capacity of all species is directly linked to these interactions consequently.

Figure 2.4: **Niche in a rock pool, located in South Devon (UK)**

Figure 2.4 shows a rock pool where a common limpet (*Patella spp.*) has an impressive growth of gutweed (*Ulva intestinalis*) on its back. The gutweed shares its fundamental niche with the calcified red algae (*Lithothamnium spp.*)—this would be the entire rock pool. They compete for space and light. However, due to herbivory on the green algae by the limpet, the realised niche remains only on the limpet's back.

Growth

The concept of population growth here can be applied to human populations and used to calculate sustainable yields of populations that are harvested, such as in fisheries management.

2.1.9 Population and Life Strategies

A population is defined as the number of organisms (individuals) of a species in a defined area. For example, the population of people in the world or the red deer population of the Isle of Rhum in Scotland. The area may be arbitrarily defined for the purposes of an investigation, for example, the population of dog whelks in a 100m² of rocky shore.

The primary factors governing a change in population size can be modelled using an input-output systems diagram.

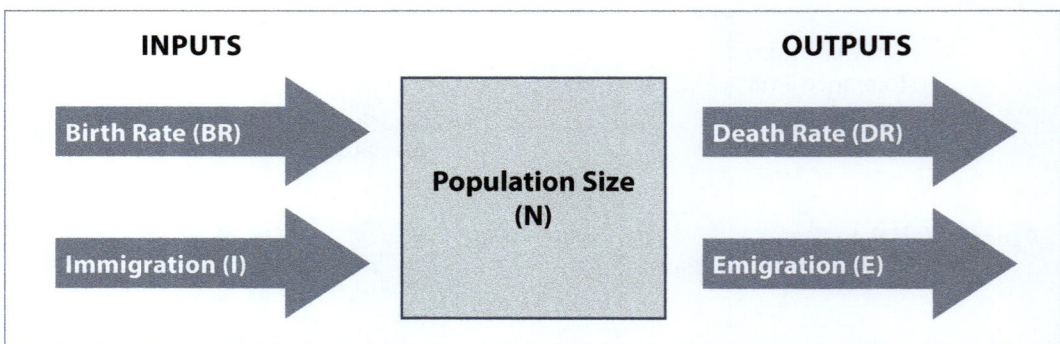

Figure 2.5: **Model of the primary factors influencing population**

Populations increase when the inputs are greater than the outputs, so population growth is calculated simply from:

Rate of Population Growth = Inputs (BR + I) – Outputs (DR + E)

This is called the growth rate; it is a measure of the time a population takes to grow, often expressed as a percentage of the population. The rate of population growth depends on a range of limiting factors that influence these primary factors.

Over time populations behave differently, with two common patterns emerging. These are shown in the typical graphs known as the S and J curves.

2.1.10 S Curves

This kind of growth pattern is shown in the graph. At the start there are few limiting factors and the population increases exponentially. This continues until the population size approaches carrying capacity. Growth rate then slows and the population tends to stabilise at carrying capacity. Here it may fluctuate, but negative feedback mechanisms regulate it close to the carrying capacity.

Typical S curve growth pattern

2.1.11 J Curves

The J curve growth pattern shows no such change in relation to carrying capacity, with the population continuing exponential growth well beyond carrying capacity before crashing back to a lower level. This is termed boom and bust population growth.

2. ECOSYSTEMS AND ECOLOGY

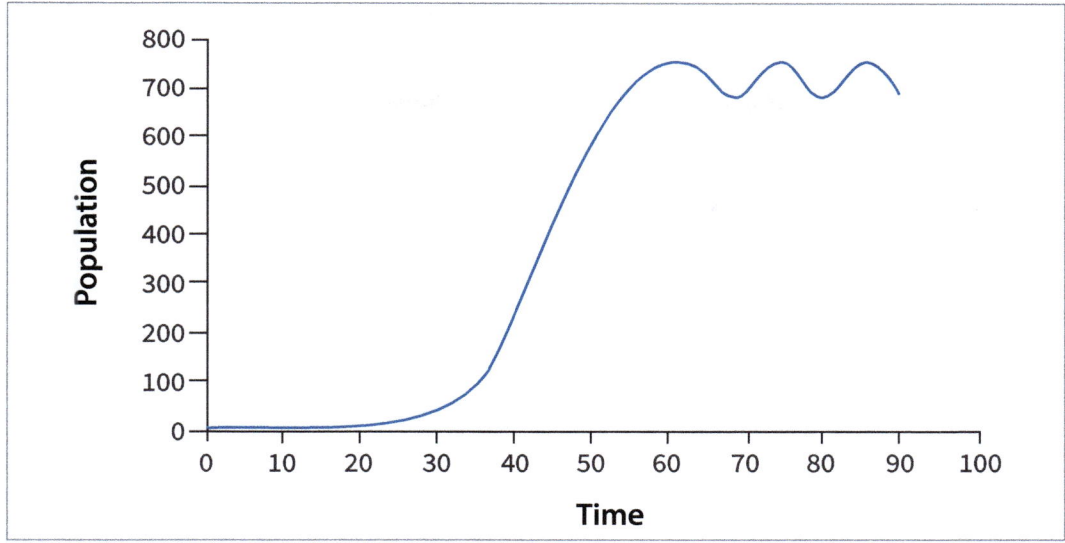

Figure 2.6: **Typical S curve growth pattern**

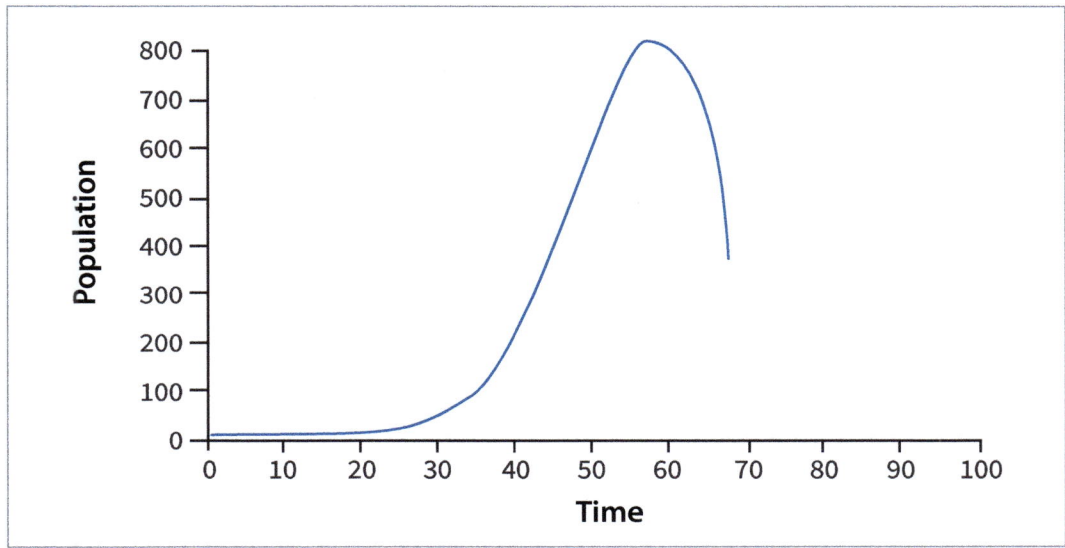

Figure 2.7: **Typical J Curve growth pattern**

2.1.12 Predator Prey Cycles

The graph in Figure 2.8 shows long term trends in the populations of Canadian lynx (*Lynx canadensis*) and snowshoe hares (*Lepus americanus*) from the Canadian tundra.

From the time marked as A until the time marked as B, you can see that an increase in hares is followed by an increase in lynx numbers. The lynx population now lags in response to change; between B and C the lynx population is increasing whilst the hares decrease due to the high levels of lynx predation. From C to D, the lynx population is in decline due to a lack of food, faster than the hare population. From D to E, the hare population now recovers as numbers of lynx are low. From E, the cycle begins again.

ENVIRONMENTAL SYSTEMS AND SOCIETIES SL

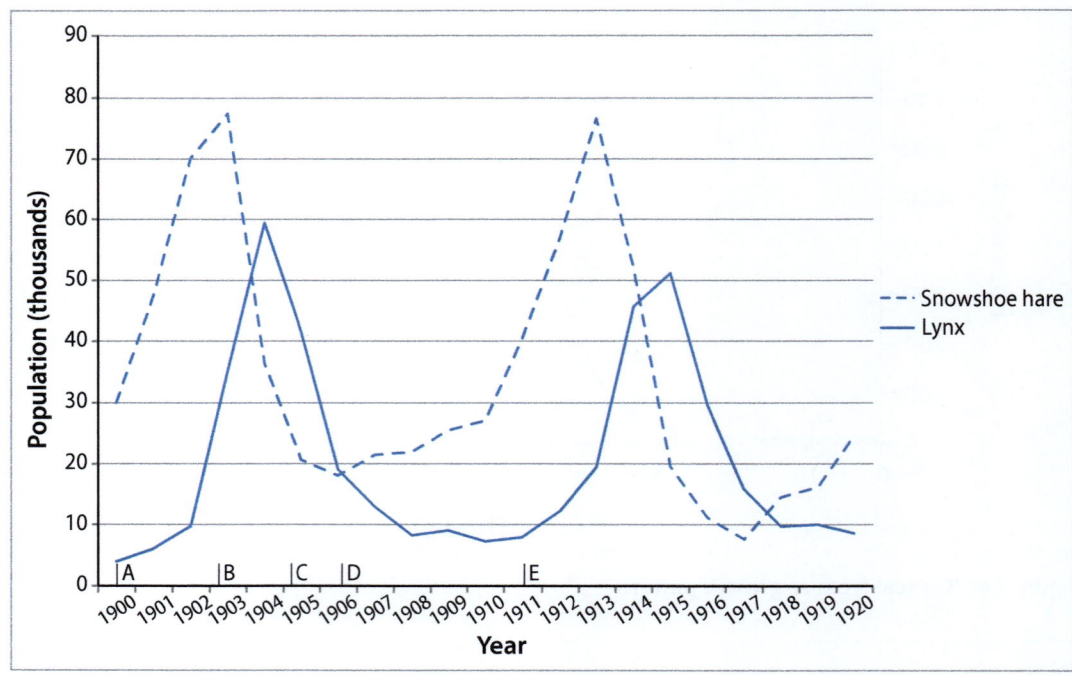

Figure 2.8: **Predator prey cycles shown by lynx and hare**

Source: Based on Hudson Bay pelt data from 1900–20. See for example: Begon, Harper, & Townsend, 2006

Practice questions: Species and Populations

1. ***Define*** the term ecology.

   ....

   ....

   ....

   ....

   ....

2. ***Define*** the term ecosystem.

   ....

   ....

   ....

   ....

3. Complete the following table:

Term	Definition	Example
Population		
	A group of organisms that interbreed and produce fertile offspring.	
		The habitat of the Indochinese tiger, *Panthera tigris corbetti*, includes a variety of types of forest.
	A species' share of habitat and resources in it. An organism's niche depends on where it lives and what it does.	

4. *Define* the following terms and give examples.

Term	Definition	Example
Competition		
Parasitism		
Mutualism		
Herbivory		
Predation		

5. *Define* the term niche.

....

....

6. *Evaluate* the niche model in 2.1.6 on page 34.

....

....

....

....

....

....

7. Figure 2.9 shows the typical relationship between host population size (*x*-axis) and parasite burden (*y*-axis).

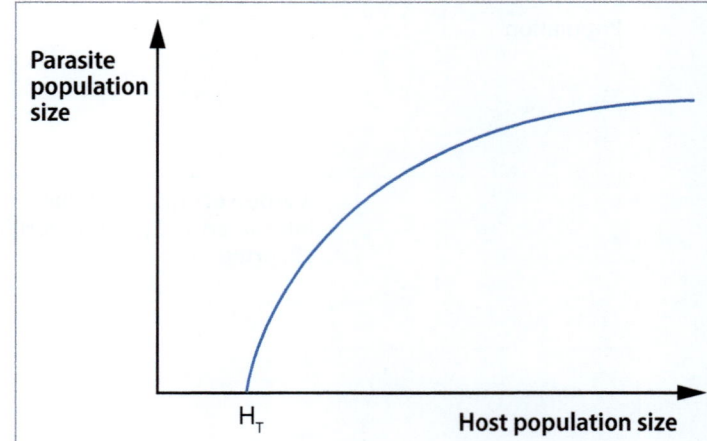

Figure 2.9: **Graph showing the average burden of parasites against population size**

Under the threshold level (H_T) the parasites are unable to establish in the population at all. *Suggest* why this is.

....

....

....

....

8. The graph in Figure 2.8 on page 38 shows long term trends in the populations of Canadian lynx (*Lynx canadensis*) and snowshoe hares (*Lepus americanus*) from the Canadian tundra.

   (a) *Identify* on the graph the start and end of a full predator and prey cycle.

   ....

   ....

   ....

   (b) *Explain* the cycle in terms of population interaction and feedbacks.

   ....

   ....

   ....

2. ECOSYSTEMS AND ECOLOGY

2.2 Communities and Ecosystems

2.2.1 Introduction to Communities and Ecosystems

You need to learn the key definitions summarised below. Importantly, note that these are living systems, and, that in both cases, they are actively interacting as they live together. Ecosystems include interactions with the non-living environments. These interactions drive energy flow and nutrient cycling through communities and ecosystems.

Community	A group of populations living and interacting with each other in a common habitat.	In the community of a South East Asian rainforest, populations of tigers interact with other species through predation and competition. Tigers may predate on mouse deer, *Tragulus napu*, and, in doing so, compete with the clouded leopard, *Neofelis nebulosa*.
Ecosystem	A community of interdependent organisms and the physical environment they inhabit.	Rainforest ecosystems have a large biomass of trees with a canopy of over 50m in height. They have high productivity despite often thin and infertile soils. The high productivity is largely due to good growing conditions all year round.

Table 2.4: **Definitions of key terms in communities and ecosystems**

2.2.2 Energy Transformation Through Photosynthesis and Respiration

These processes are responsible for the majority of energy transformations of ecosystems and communities. They are also key elements in the carbon cycle. Note that in terms of energy and carbon, one is the opposite of the other. As the energy requirements of living things are based on these two processes, they also drive other material (nutrient) cycles such as the nitrogen cycle.

2.2.3 Photosynthesis

The significance of the primary producers is great, as without them nothing else would exist. The process of photosynthesis is carried out by producers such as green plants and algae. The green pigment chlorophyll is needed for the process to take place. Photosynthesis is a complex series of reactions, but it can be simplified into the following equation:

Photosynthesis

$$\text{carbon dioxide} + \text{water} \xrightarrow[\text{light}]{\text{chlorophyl}} \text{glucose} + \text{oxygen}$$

As well as plentiful supplies of carbon dioxide and water, this process needs sunlight as the energy supply and warm temperatures enable it to work rapidly.

2.2.4 Transformations of Energy and Matter

Matter transformations occur as water as a liquid produces oxygen gas and the carbon is fixed from carbon dioxide gas to a solid form as glucose. Glucose can then be used to make other organic compounds, including carbohydrates, proteins, and lipids. Thus, it is the building block of all living biomass. Energy transformations occur as light energy is now converted into chemical energy. Some heat is lost during the process. These processes can be summarised in a systems diagram such as Figure 2.10.

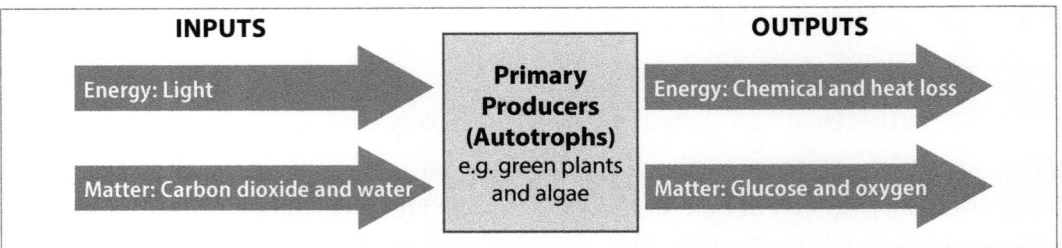

Figure 2.10: **Transformations of energy and matter in respiration**

2.2.5 Respiration

All living organisms need to make use of the energy available in organic matter. Producers use the material they have made themselves. Consumers ingest the bodies of other organisms to get this energy. Decomposers break down organic matter in the dead organic matter pool of the system.

The basic equation for this release of useful energy, known as respiration, is:

Respiration

$$\text{glucose} + \text{oxygen} \underset{mitochondria}{\Rightarrow} \text{carbon dioxide} + \text{water} + \text{heat}$$

All organisms need to carry out this process all the time to survive; it provides energy for all living processes.

2.2.6 Transformations of Energy and Matter

Matter transformations occur as stored chemical energy such as glucose produces carbon dioxide gas and liquid water.

Energy transformations occur as the stored chemical energy in glucose is now converted into a more accessible chemical energy that is readily available for processes such as movement when needed. Some heat is lost during the process.

These processes can be summarised in a systems diagram such as Figure 2.11.

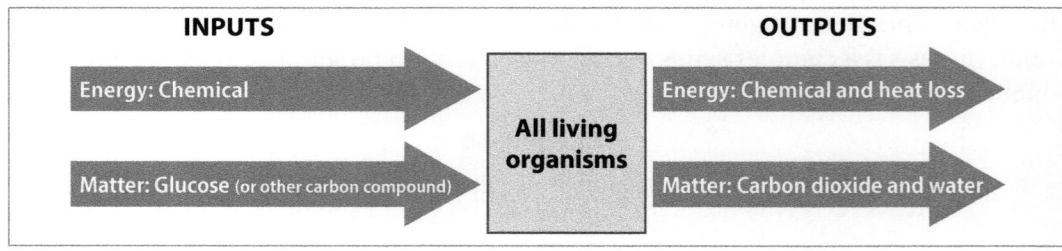

Figure 2.11: **Transformations of energy and matter in respiration**

2.2.7 Entropy Changes Due to Respiration and Photosynthesis

In accordance with the laws of thermodynamics, ultimately all the energy fixed from light by photosynthesis will be lost as heat, largely due to respiration. Therefore, whilst individual organisms can maintain negative entropy (that is, increased order) with their chemical organisation, the overall system increases in entropy due to respiration.

Overall order is maintained due to the constant input of light. But if photosynthesis stops due to an absence of light, then moving to greater disorder would be proven as all organisms would die and decompose (entropy law).

2. ECOSYSTEMS AND ECOLOGY

2.2.8 Trophic Structure: Biotic Components

The biotic components of an ecosystem interact through their feeding relations. In ecology the term 'trophic' is used to describe a feeding relationship.

Term	Definition	Example
Trophic level	The position that an organism occupies in a food chain, or a group of organisms in a community that occupy the same position in food chains.	Herbivore
Producers	Producers are also called 'autotrophs', meaning self-feeding. In most ecosystems these are green plants or algae, photosynthetic organisms that form the base of the food chain. In some deep-sea communities with no light, there are some chemosynthetic organisms—bacteria that feed on chemicals.	Lowland oak *Quercus robur*
Consumers	Any organism that eats or gains nutrition from another. They are also heterotrophs, which means they feed on another organism.	Any from the list below
Herbivores	Heterotrophs that are also termed primary consumers. They consume primary producers. 'Herbivore' literally means 'eats grass'.	Cottontail rabbits *Sylvilagus floridanus*
Carnivores	Heterotrophs that are secondary consumers or greater. They divide into first order, second order, third order carnivores, and so on, up to a top carnivore at the end of the food chain. 'Carnivore' literally means 'eats meat'.	Bluefin tuna *Thunnus thynnus*
Omnivores	These are heterotrophs that feed at any trophic level. They often have a varied diet. 'Omnivore' literally means 'eats all'.	Badger *Meles meles*
Detritivores	Heterotrophic organisms (normally animals) that consume dead organic matter by ingestion. 'Detritivore' literally means 'eats detritus'.	Earthworm *Lumbricus terrestris*
Decomposers	These organisms are fungi and bacteria that break food down outside their bodies, by secreting enzymes into the environment. As this process is inefficient, they are important in recycling nutrients. Note they don't ingest (eat) as animals do.	Penny bun fungus (cep) *Boletus edulis*

> **Examples**
> When possible, it is better to use local-named examples from an ecosystem you have studied.

Table 2.5: **Definitions of biotic components in the ecosystems**

2.2.9 Comparison of Terms Used for Biotic Components

Biotic components are further divided into groups in a number of different ways, as summarised and defined in Table 2.6.

1st Trophic level	Autotroph	Primary producer	
2nd Trophic level		Primary consumer	Herbivore
3rd Trophic level		Secondary consumer	1st Order carnivore
4th Trophic level		Tertiary consumer (end of chain)	2nd Order carnivore
5th Trophic level	Heterotroph		Top carnivore
		Omnivore	
Decomposer subsystem		Detritivore	
		Decomposer	

Table 2.6: **Terms used for biotic components**

Practice questions: Communities and Ecosystems 1

1. Complete the following table:

Term	Definition	Example
Ecosystem		
Community		

2. Complete the table listing the chemical (matter) input and outputs of both photosynthesis and respiration.

Photosynthesis		Respiration	
Inputs	Outputs	Inputs	Outputs

3. *Draw* diagrams to show all the energy transfers that take place in both photosynthesis and respiration.

4. *Describe* the roles of the following trophic levels in the ecosystem:
    (a) Producer
    (b) Consumer
    (c) Decomposer

....

....

....

....

5. Complete the table by giving numbers for the trophic levels in the first column. Complete the rest of the table using the following terms.

Heterotroph, Decomposer, 2nd order carnivore, Herbivore, Primary producer, Secondary consumer, 1st order carnivore

	Autotroph		
		Primary consumer	
		Tertiary consumer	
			Top carnivore
Decomposer subsystem		Detritivore	

2.2.10 Trophic Structure Through Food Chains, Webs, and Pyramids

All three systems are used to consider the feeding structure of communities. As you revise each, consider the use of each approach to understanding the feeding structure in different contexts.

2.2.11 Food Chains

These show the sequence of organisms in successive trophic levels within a community. Food chains are flow diagrams that show feeding relationships and, therefore, the movement of matter and energy. Some examples of food chains are shown in the following figures.

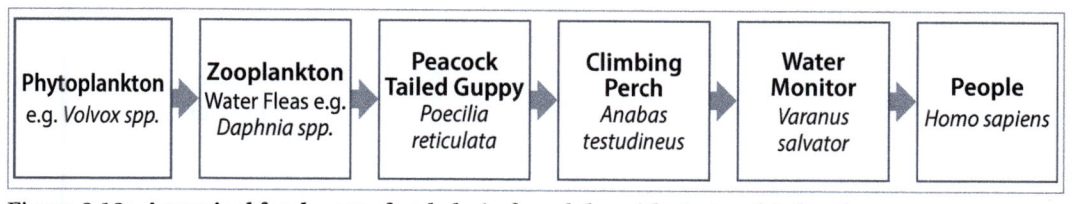

Figure 2.12: **A tropical freshwater food chain for a lake with six trophic levels**

Note the abbreviation **spp.** refers to a *number* of possible species, **sp.** refers to an unknown, *singular* species of that genus.

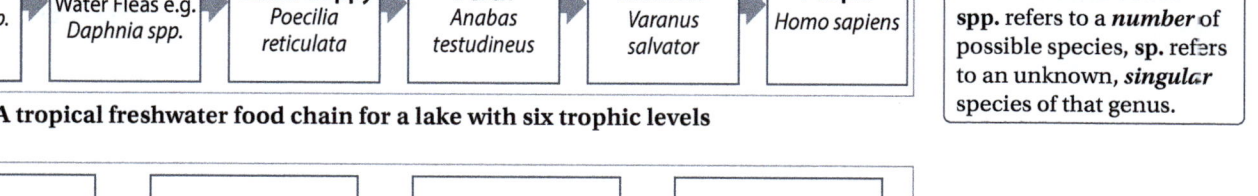

Figure 2.13: **South East Asia forest food chain with four trophic levels**

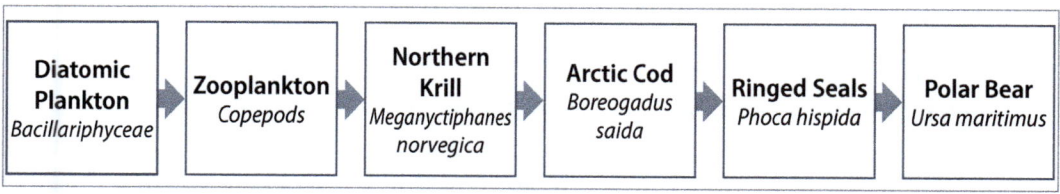

Figure 2.14: **An Arctic food chain with six trophic levels**

2.2.12 Application and Limitation of Food Chains

Food chains as simple as these examples are rare in nature, even in systems with low diversity of species. In the Arctic food chain in Figure 2.14, there is alternate prey for each species feeding on a lower trophic level. The polar bear can choose from one of six species of seal, and may try to eat other mammals like walrus or people if it is really hungry.

Food chains are useful for analysing aspects of ecosystem function such as the response to persistent pollutants that pass on through food chains. They are also useful to study diseases that follow food chains, such as bovine spongiform encephalopathy (BSE, or mad cow disease).

Disease and food chains
Not many infectious diseases pass down food chains. Examples like mad cow disease are unusual. Many viruses and bacteria are species specific.

2.2.13 Bioaccumulation and Biomagnification

Persistent pollutants, DDT (1.5.3) and mercury for example, are said to bioaccumulate. This means that they accumulate in the tissues of living organisms as they cannot be broken down, so, consequently, they accumulate in the trophic level.

These pollutants may then biomagnify along the food chain, as they increase in concentration with each trophic level. See the example from Minamata Bay for further details (1.5.5).

Note: this process does not happen with disease, only persistent pollutants such as heavy metals and pesticides. Only a few diseases can actually pass along a food chain; most cannot move from one species to another.

2.2.14 Food Webs

Food webs show more complex and complete feeding patterns than food chains. You should learn one example of a food web like the one in Figure 2.15, ideally from a local case study.

Human Impacts
You need to be able to relate an understanding of topic 2 to human impacts on the environment and food production. Consider possible applications of theory as you revise this topic.

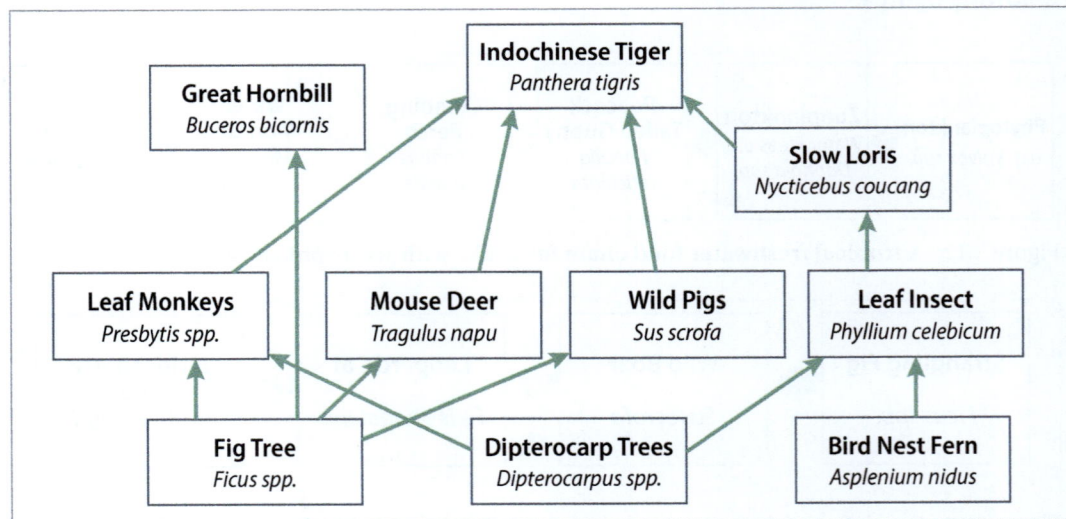

Figure 2.15: **South East Asian forest food web**

2.2.15 Application and Limitation of Food Webs

Although food webs show complex feeding relationships, they are still not complete. They do not clearly show the differences in the quantity of living organisms in the different trophic levels. Detailed information on the feeding relationships of a wide variety of species is hard to come by. Species may even change their position during their life span.

Each species would need a close examination of its diet, through field observations, or dissection of faeces or gut contents. Studies like this can be very useful to examine impact on feeding relations in a food web. For example, the impact of invasive species on native species through competition in a food web (see the case study on otter vs. mink in section 3.4.8).

2.2.16 Pyramids

Pyramids show how ecosystems handle transfers of energy and matter between trophic levels.

Pyramids are graphical models of communities, drawn like bar graphs on their side, with a central y-axis about which the pyramid is symmetrical. They show the total quantity in terms of numbers, biomass, or energy.

2.2.17 Pyramid of Numbers

The simplest kind of pyramid presents count data of populations in each trophic level.

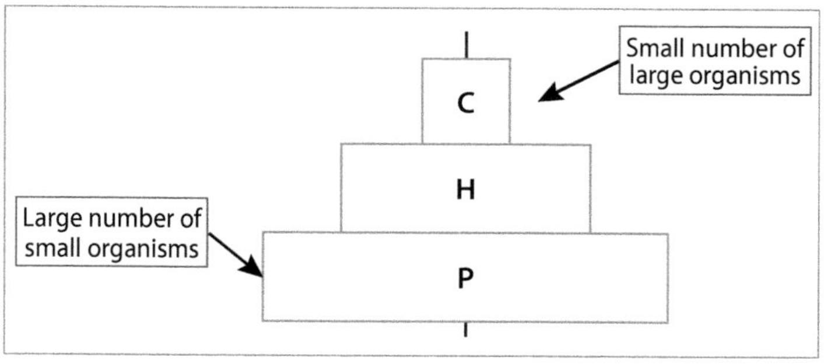

Figure 2.16: **An upright pyramid of numbers**

As an example, this could be the first three trophic levels from an African savanna. Here grazing zebra and gazelle are preyed on by lions and hyenas. There is a steady decline in numbers from producer (P), to herbivore (H), to carnivore (C), hence this typical upright pyramid narrows towards its apex.

2.2.18 Ecological Efficiency and the Second Law of Thermodynamics

There is a tendency for the available energy to decline with each transfer, due to the second law of thermodynamics. Heat is lost at every link in the chain. Biomass and numbers, therefore, fall along the food chain. We may expect pyramids to show the '10% rule', which suggests that only around 10% of the energy in one trophic level will pass to the next. Use this equation to measure the efficiency of transfer (ecological efficiency) and test the 10% rule for a given system:

$$\text{Ecological efficiency} = \left(\frac{\text{Amount in the higher trophic level}}{\text{Amount in the lower trophic level}} \right) \times 100\%$$

Real data shows that this efficiency of transfer is rarely 10%; in some instances it is much lower and in others far greater. The rule is still used to indicate a large loss of energy between trophic levels, resulting in a typical upright pyramid. Longer food chains suggest better efficiency, as is typical in aquatic ecosystems.

2.2.19 Inverted Pyramid of Numbers

Pyramids of numbers sometimes show exceptions to the 10% rule and can even be inverted. Although consumers are often much bigger than producers, it is not always true. Both herbivores and carnivores may be numerous and small, even though they are higher up the food chain.

This example from the African savanna now has a fourth trophic level showing the parasite burden on the lions. Parasites such as fleas, mosquitoes, or worms exist in very large numbers for every individual lion.

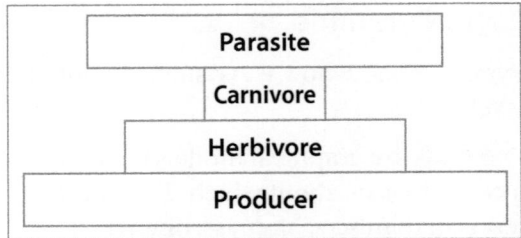

Figure 2.17: **An inverted pyramid of numbers**

2.2.20 Pyramids of Biomass (Standing Crop)

The total number of organisms can be counted and weighed. Then this data can be used to make pyramids of biomass in each trophic level, removing the problem of organism size.

However, even pyramids of biomass are sometimes inverted. Here is an example of using data from a temperate marine ecosystem.

Organism	Biomass (dry mass) in g/m²
Phytoplankton (P)	4.0
Zooplankton (H)	21.0

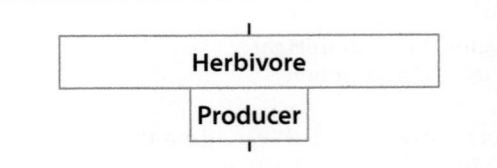

Figure 2.18: **Data and pyramid of biomass from the English Channel**

There are two reasons why this kind of pyramid can be found:

1. Biomass is not equivalent to energy.
2. The pyramid shows standing crop biomass, not energy flow (see pyramids of standing crop).

2.2.21 Pyramids of Energy

To get over this problem of inversion, then, we should estimate the energy in each trophic level. Energy can be calculated per gram by controlled combustion of biomass to heat water. The energy change can be calculated by the temperature change of the water. Total energy in the trophic level can then be estimated. These pyramids can still be inverted, though, if the units involve only standing crop and no energy flow.

2.2.22 Pyramids of Standing Crop

Pyramid diagrams may show the fixed quantity of numbers, biomass, or energy that exists at a particular time in a given area, or averaged from many of these measurements. This is termed 'standing crop'. There is no element of time in this measure; it is rather like a still photograph. The units would be number, dry biomass, or energy kg/m², or J/m³. It is still possible for these standing crop pyramids to be inverted. For example, sheep grazing a field may have a greater biomass averaged over the year than the grass in that field.

2.2.23 Pyramids of Productivity

Productivity is the rate of change in the biomass or energy within a community, trophic level, or individual. It is normally expressed, therefore, as either dry mass units or joules of energy, as a rate. The unit must be against time, normally combined with a unit of area or volume such as $kg/m^2/year$ or $J/m^3/week$. Effectively this is more like taking a video than a photograph. The sheep in the field cannot increase their mass faster than the rate at which the grass grows. Pyramids of productivity are never inverted.

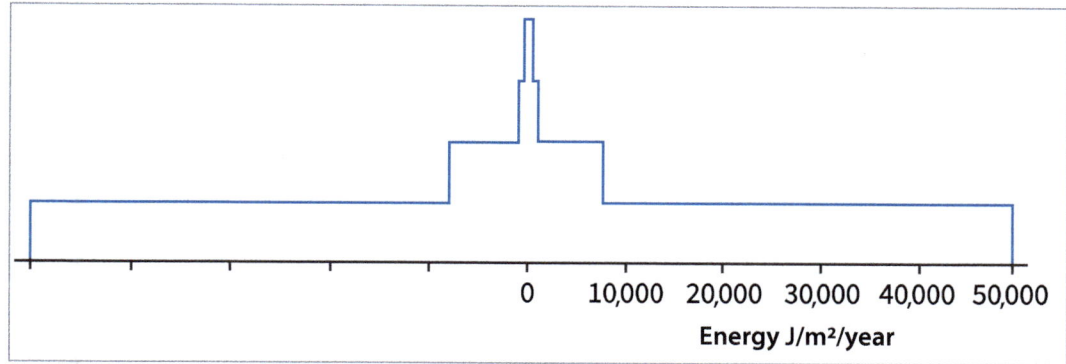

Figure 2.19: **A pyramid of productivity of a temperate freshwater stream**
Source: Data from Odum, 1957

2.2.24 Comparing Standing Crop and Productivity Pyramids

Pyramids can also be drawn as energy flow models, with boxes and pipes. This model shows an inverted pyramid of the North Sea marine ecosystem. Notice that the pyramid is only inverted in terms of standing crop, not productivity.

- Boxes (dark blue) represent standing crop in $105 J/m^2$
- Pipes (light blue) represent productivity (energy flow) in $105 J/m^2/week$

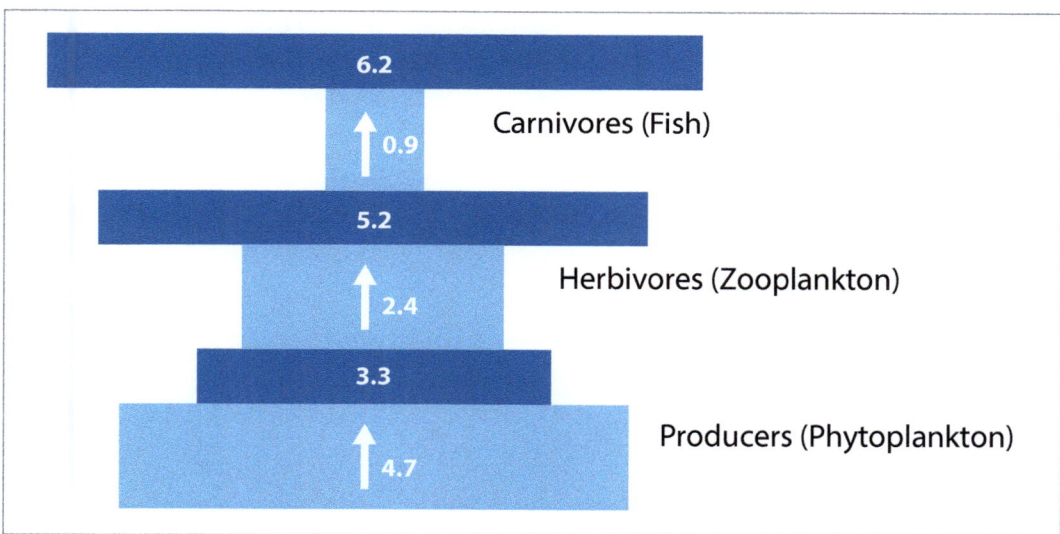

Figure 2.20: **Energy flow model of the North Sea**

Practice questions: Communities and Ecosystems 2

1. *Identify* the following, using the diagrams of food chains and food webs (Figure 2.12 to Figure 2.15):

   (a) A tertiary consumer from a tropical aquatic system

   ....

   (b) A primary consumer from the Arctic

   ....

   (c) Two primary producers in South East Asian forest

   ....

   (d) A top carnivore from the Arctic

   ....

   (e) A forest herbivore

   ....

   (f) A marine carnivore

   ....

   (g) A species occupying the fifth trophic level in the Arctic

   ....

2. *Draw* a food chain and a food web (with at least six species) from an ecosystem that you have studied.

3. *Explain* a food chain in terms of the two laws of thermodynamics. *Estimate* the percentage of energy found in the tertiary consumers of any food chain in comparison with the producers.

....

....

....

....

4. *Explain* why food chains longer than six trophic levels are rarely found.

....

....

....

5. Complete the following table, using the appropriate SI units, for each of the six types of pyramid.

Pyramid type	Standing crop	Productivity
Numbers		
Biomass		
Energy		

6. From the above table, *identify* which kind of pyramid is least likely to fit the 10% rule and *explain* why this is.

....

....

....

....

7. *Identify* which pyramid could never be inverted and *explain* why this is.

....

....

....

Questions 8–14

Use the data table below for questions 8–14. Students collected this data on invertebrates from a freshwater lake. They divided them up into four trophic levels according to both research and observations.

Invertebrate Group	Number of Individuals	Average Dry Biomass (mg ±0.01)	Trophic level	Numbers			
				D	H	C1	C2
Flatworm	10	0.40	d c1	5		5	
True worms	37	0.70	d	37			
Leeches	23	1.60	c1			23	
Ramshorn snails	57	0.80	h		57		
Pond snails	29	3.50	h		29		
Valve snails	30	0.21	h		30		
Pea cockles	31	1.20	D	31			
Water fleas	636	0.01	D H	318	318		
Ostracods	321	0.05	D H C	107	107	107	
Water hoglouse	24	1.20	D	24			
Shrimp	132	0.40	D	132			
Water mites	26	0.30	D H	13	13		
Lesser boatmen	34	4.40	D H	17	17		

(D = detritivore, H = herbivore, C1 = first order carnivore and C2 = second order carnivore)

8. ***Draw*** a pyramid of numbers for the pond ecosystem using the steps (a), (b) and (c). Add the detritivores in a block underneath the pyramid.

    (a) Total the numbers for each trophic level, then draw an x-axis for the total amount and y-axis on the midway point.

    (b) Next divide the total for a trophic level in half, and use these values to plot a set of steps as shown.

    (c) Add the mirror image of these steps and give a suitable scale to the x-axis and labels for the y-axis.

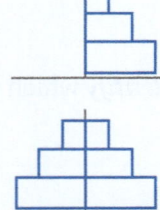

9. Calculate the biomass for each species in each trophic level by multiplying the individual biomass by the number. Total this to give biomass for each trophic level.

   ....

   ....

   ....

   ....

10. Use the biomass data to *draw* a pyramid of biomass.

11. *Compare* your two pyramids and explain any differences.

    ....

    ....

    ....

    ....

12. *List* observations that would help the students decide on the trophic level.

    ....

    ....

    ....

    ....

13. Some groups have more than one trophic level. *Describe* how has this data been manipulated to represent this.

   ....

   ....

   ....

   ....

14. *Calculate* the ecological efficiencies for predators and prey in the pyramid of numbers and the pyramid of biomass from the freshwater lake.

	Prey (Herbivores and Detritivores)	Predator (1st and 2nd order carnivores)	Efficiency (predator/prey) × 100
Numbers			
Biomass			

    Now combine the totals for the herbivores and detritivores as the carnivores eat both. Next combine the carnivores into one total. Complete the table and calculate the efficiency values from this data only.

2.3 Flows of Energy and Matter

Energy flows, matter cycles

Energy flows through the ecosystem; when it is used, it is converted to a new form. Energy cannot be recycled. Matter cycles through the ecosystem. The amount of matter is fixed (finite) and is constantly broken down and built up into new forms.

2.3.1 Ecosystem Components

The components of an ecosystem can be first divided into the abiotic, or physical, components and the biotic, or living, components. Look for these components in the ecosystem diagram (Figure 2.21). Here they are divided by a dotted line running diagonally across the diagram. The biotic components in the diagram are divided by a vertical dotted line showing the division between the two key subsystems of the grazing and decomposer food chains.

2.3.2 Energy Flows and Matter Movements

Ecosystem components are connected by flows of energy and matter. Ultimately the sun supplies all energy to the vast majority of ecosystems; note that all the energy flows are in a linear pathway, which is shown in yellow in Figure 2.21 as it moves from light to chemical to heat through the system. Movements of complex biochemicals include usable energy, so here the energy is alongside the blue lines, which indicates the movement of materials through the system. Note how the matter cycles return nutrients to the producers in the food chain.

2. ECOSYSTEMS AND ECOLOGY

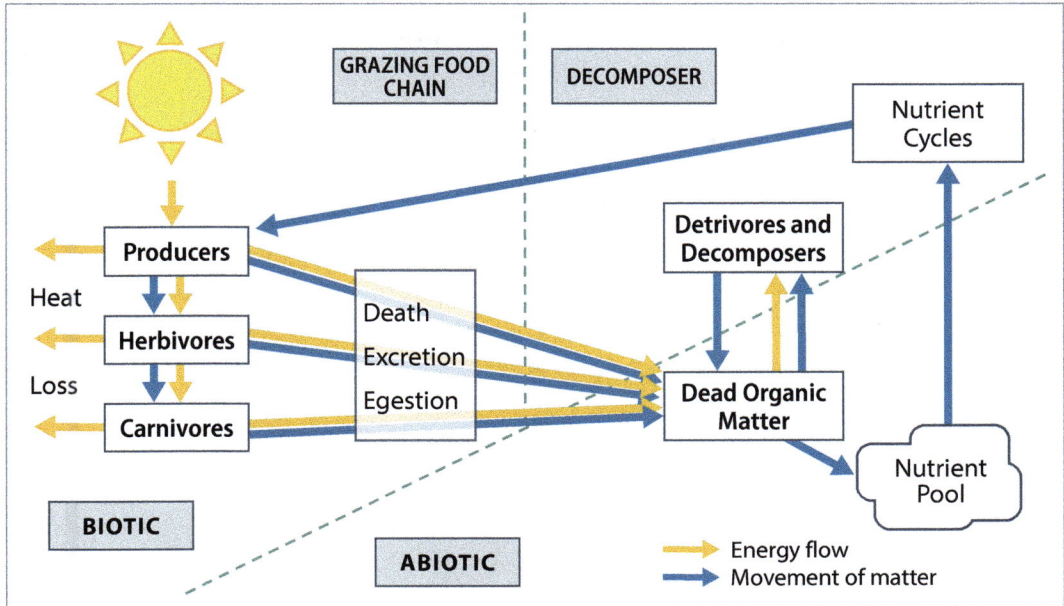

Figure 2.21: **An overview of structure in a generalised ecosystem**

2.3.3 The Fate of Solar Radiation: Insolation in the Atmosphere

Considerable amounts of energy are unavailable to ecosystems as they are reflected or absorbed by inorganic materials, including clouds and dust particles in the atmosphere. The amount of light lost in this way varies considerably due to location, seasonal change, weather, and pollution, but may be around 50% of the insolation.

> **Insolation**
> This refers to the total radiation energy arriving at the Earth from the sun.

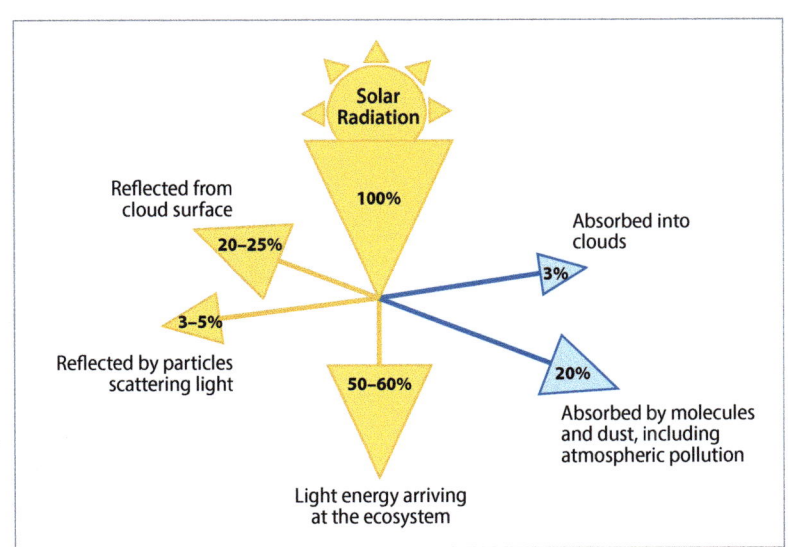

Figure 2.22: **Energy flow diagram of incoming solar radiation in the atmosphere**

2.3.4 Energy Flow Diagrams Show Transfer and Transformation of Energy

You should be able to recognise the transfers and transformations that energy goes through on its journey through the Earth's ecosystems, such as the ones on the previous page. Energy flow through the entire community can be represented using energy flow diagrams that show the storage of energy in component boxes and flows of energy through arrows. The flow arrows sometimes have a width proportional to the amount of energy flowing through that pathway.

2.3.5 Productivity and its Calculations

Productivity is the rate of growth of an organism, population, community, or trophic level. It is the production of the defined group, over time. It can be measured with a variety of units, but biomass is quite common.

Gross production (GP) is that which occurs before respiration (R), and net productivity (NP) is after. Therefore:

$$GP - R = NP$$

Primary productivity is the rate of change in producers, secondary the rate of change in consumers.

You need to know how to calculate values for all of these.

> **Growth**
> The concept of growth links to topic 2 and to other topics. Living organisms grow when they gain dry mass. Net productivity is the growth of an organism and is also used to indicate sustainable yield (8.2.3).

2.3.6 Insolation in the Ecosystem

Figure 2.23 shows what happens to insolation in the ecosystem. The efficiency of transfer between sunlight and chemical energy in the plant is very low, between 2–5%.

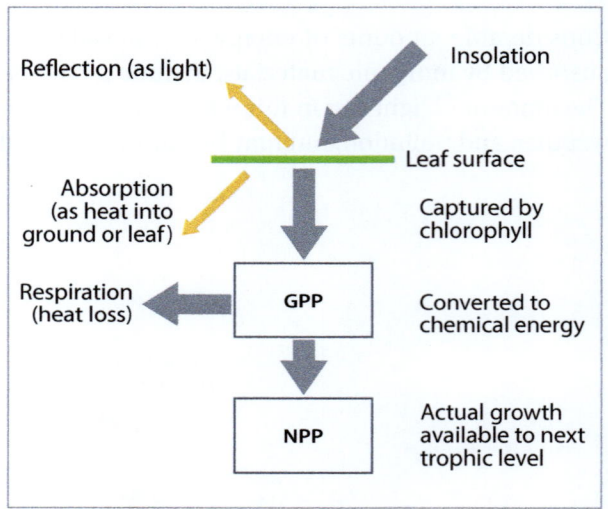

Figure 2.23: **Energy flow diagram of incoming solar radiation in the ecosystem**

- Radiation lost through reflection and absorption and re-radiation of heat
- Light energy converted to chemical
- Loss of heat from plant respiration
- Production of NPP (Net Primary Productivity), from GPP (Gross Primary Productivity)
- Later efficiency of chemical energy transfer along the food chain are higher than this— often around 10–20%

2.3.7 Calculations of Primary Productivity

Gross Primary Productivity (GPP) is the amount of photosynthesis carried out by the producers; it indicates the total conversion of light to chemical energy during photosynthesis. Net Primary Productivity (NPP) is the actual growth of the producer, the part of GPP that is available after the producer has carried out respiration (R).

GPP, NPP, and R are all linked by the equation:

$$NPP = GPP - R$$

2.3.8 Calculations of Secondary Productivity

Any input of food will move down one of the three pathways shown in the diagram. Gross Production, in reality, is equivalent to food assimilated, and so it is equal to the food ingested minus faeces lost (this food never properly entered the body's tissues).

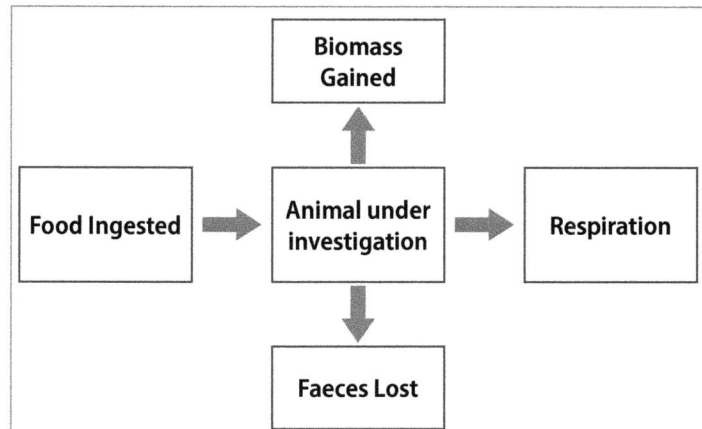

Figure 2.24: **Simplified model to show secondary productivity**

How would you calculate NP from the model?

Respiration is not easy to measure directly, and so if your answer to this question is Food Ingested-Faeces Lost-R, then this is correct, but it may not be possible in practice. Again, the actual change in biomass is the Net Secondary Productivity (NSP).

The Gross Secondary Productivity (GSP) cannot be measured directly, but it is calculated from this equation:

> GSP = Food Eaten – Faecal Loss

The NSP value is then used to calculate the respiration from this equation:

> NSP = GSP – R

2.3.9 The Cycling of Materials

Matter is transformed from one form to another and transferred from one place to another by a variety of living and non-living processes.

The carbon, nitrogen, and water cycles are all material cycles you need to know.

ENVIRONMENTAL SYSTEMS AND SOCIETIES SL

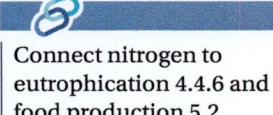

Connect nitrogen to eutrophication 4.4.6 and food production 5.2.

2.3.10 The Nitrogen Cycle

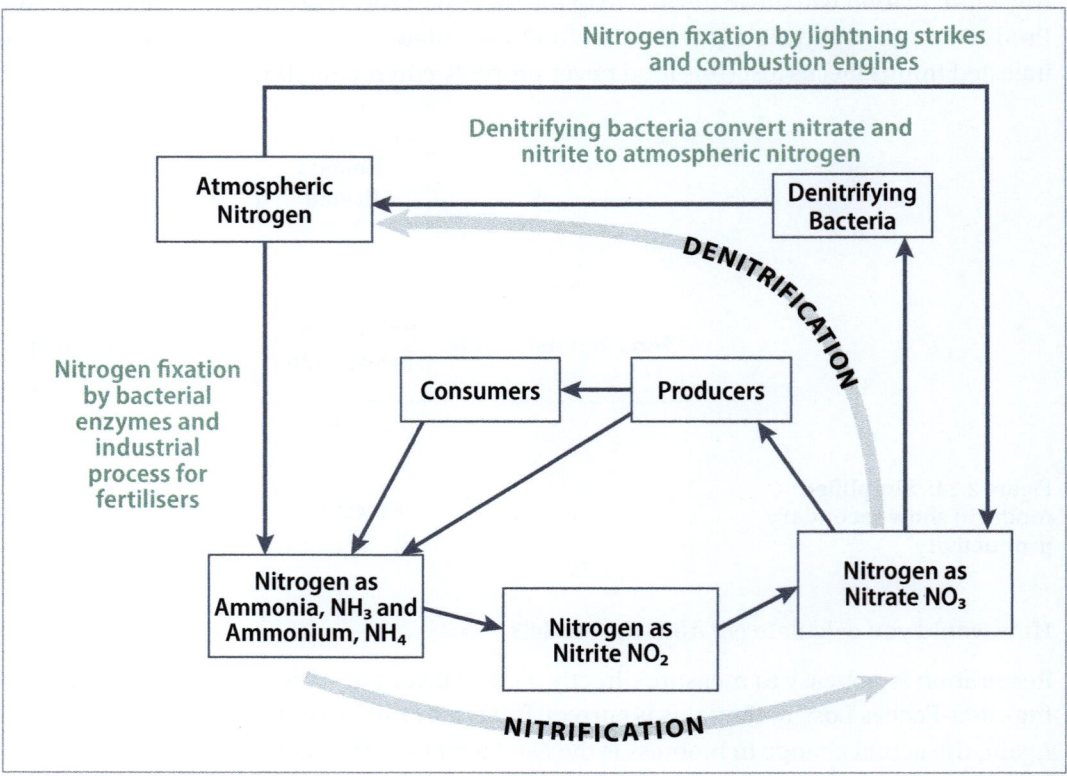

Figure 2.25: **The nitrogen cycle**

- Nitrogen fixation by blue green algae and bacteria converts nitrogen to ammonia (transformation).
- Nitrogen fixation by lightning strikes and fertiliser production direct to nitrate (transformation).
- Nitrification is an oxygen demanding process carried out by bacteria (transfer).
- Denitrification removes nitrogen stores from the soil and returns them to the atmosphere, carried out by bacteria in water-logged conditions (transformation).
- Nitrogen is absorbed as nitrate by plant roots from the soil (transfer).
- Nitrogen passes down the food chain as protein in tissues (transfer).
- Nitrogen is removed from the food chain by death, egestion, and excretion (transfer).
- Nitrogen is abundant in the atmosphere (78% of atmosphere).
- Nitrogen fixation carried out for fertiliser production is essential for the global food supply (transformation).
- Nitrification causes high BOD in water (4.4.3).
- Small amounts of nitrogen can be found in fossil fuels.

2.3.11 The Carbon Cycle

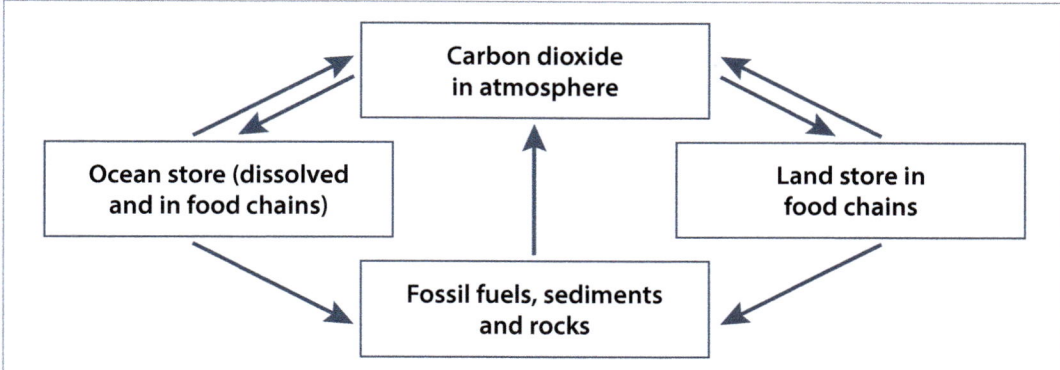

Figure 2.26: **The carbon cycle**

- Carbon moves from the atmosphere to living organisms by **photosynthesis** (transformation).
- Carbon moves into the atmosphere by respiration, decomposition, and combustion (transformation).
- Carbon dioxide dissolves in sea water as inorganic carbon store, for example, carbon compounds in deep sea mud.
- Carbon stored in ecosystem biomass as organic carbon, for example, protein in plankton and fish.
- Carbon moves along the food chain due to feeding activity through all biomass (transfer).
- Inorganic carbon sediments (limestone, chalk, shale, and coal) form stores as rocks and minerals (transfer).

Connect the carbon cycle to Figure 1.3 and climate change (7.2).

Nitrogen and carbon cycles

You need to learn the carbon and nitrogen cycles and be able to draw out a diagram, including the key processes bullet pointed under the diagrams. You may also be asked to construct a systems diagram from given data or interpret a given diagram, including human impacts on the cycles.

Practice questions: Flows of Energy and Matter

1. *Outline* the differences in the way ecosystems handle energy and matter.

   ....

   ....

   ....

   ....

2. *Explain* why ecosystems handle energy and matter differently using the laws of thermodynamics.

   ....

   ....

   ....

   ....

ENVIRONMENTAL SYSTEMS AND SOCIETIES SL

	Insolation	Producers	Herbivores	Carnivores
Efficiency		5%	10%	10%
	MJ/m²/year	MJ/m²/year	MJ/m²/year	MJ/m²/year
Spitsbergen	2,500	125		
London	3,200	160		
Miami	6,500	325		

3. **Calculate** the energy available for herbivores and carnivores (assume the 10% rule), then complete the table above.

   ....

   ....

   ....

4. **Suggest** why is so little of the incoming solar radiation is converted into energy by the producers.

   ....

   ....

   ....

   ....

5. Animals in colder climates tend to be larger than hotter ones to help conserve body heat. Use this fact and the above calculation to **explain** why one polar bear needs 50 km² of tundra but the Malaysian sun bear needs 15 km² of rainforest.

   ....

   ....

   ....

   ....

6. Write out the full name and **define** the following terms:

	Full name	Definition
GP		
NP		
R		
NPP		
GPP		
GSP		
NSP		

2. ECOSYSTEMS AND ECOLOGY

7. **Calculate** NPP, R and GPP using the provided data from the light and dark bottle techniques carried out on a temperate freshwater lake. Use the average NPP for all five bottles and give your answers in mg/m³/hr.

Bottle Type and Depth (m)	Oxygen (mg/m³/hr)
(Light – Photosynthesis + Respiration) 0.1	612
(Light – Photosynthesis + Respiration) 0.5	350
(Light – Photosynthesis + Respiration) 1.0	1049
(Light – Photosynthesis + Respiration) 1.5	1166
(Light – Photosynthesis + Respiration) 2.0	1633
(Dark bottle—Respiration only) 2.0	–2245

Some IB students were provided with seven giant pond snails and a water tank. The snails were weighed during class times and the faeces filtered off along with uneaten food. They were provided with pond weed, which was weighed. The water was changed at the same time. Their results from the laboratory measurements are shown below.

DATES	Total Wet Mass of Seven Snails (g ± 0.05)	DATES	Food Given (g ± 0.05)	4th May	
				waste food (g ± 0.05)	faeces produced (g ± 0.05)
27th April	380.6	28th April	18.8	2.0	8.3
4th May	390.8	29th April	219.2		
		TOTAL	268.9		

A laboratory estimation of secondary productivity of the golden apple snail, Pomacea canaliculata over eight days

8. **Calculate** NSP, R, and GSP from the data using the following steps:

    (a) **Calculate** NSP in g/day of wet mass for all seven snails during the eight days.

    ....

    ....

    (b) **Estimate** the amount of food assimilated. Combine the mass of waste and faeces and remove this from the total food given. Convert this figure to g/day to give GSPP.

    ....

    ....

    (c) Now **calculate** R.

    ....

    ....

9. Use your understanding of ecosystem structure and function to *explain* how it is possible that matter from a *Tyrannosaurus rex* dinosaur could be present in your own body tissues but the energy contained in the dinosaur's body could not.

   ....

   ....

   ....

   ....

10. *Explain* how atmospheric nitrogen can end up as protein in a rabbit's lung.

    ....

    ....

    ....

    ....

11. *State* the energy source that drives the carbon cycle and *identify* all of the processes that transfer and transform carbon.

    ....

    ....

    ....

    ....

2.4 Biomes, Zonation, and Succession

2.4.1 Biomes and Their Distribution

A biome is a large-scale biogeographical region composed of similar species of animal and plant, in a similar climatic region. Many are visible from space, where the differing characteristics of the biome are shown by the colours.

Biomes can be divided into five major classes: the tundra, desert, forest, grassland, and aquatic. The classes of biomes have numerous subdivisions. You need to revise the following biomes in depth, and you need to be able to explain the distribution, structure, and relative productivity of these biomes.

The general pattern of biome distribution is linked to latitude (due to variations in total insolation) but also due to the patterns of atmospheric circulation.

You should be able to describe and explain the distribution of the biomes in Tables 2.7 and 2.8 on the next pages.

Biome examples:

Choose at least four contrasting pairs of biomes to revise for the exams, for example, rainforest and grassland could be one pair. You can use the tables 2.7 and 2.8 to review details for your examples.

2.4.2 The Tricellular Model of Atmospheric Circulation

Between the equator and the poles in both hemispheres, there are three large convection cells that circulate the air in the atmosphere and move heat energy from the equator to the poles. The cells have a powerful influence on biome distributions due to their influence on temperature and precipitation. For example, the desert belts are located under the cool, dry air falling down from the tropical cell and the rainforest under the hot, humid rising air that produces heavy rainfall.

See it for real:
Search on the internet for a good quality photographic image of the Earth's surface from space. Look for the biome types. How many major types can you recognise? The dry deserts are shown in sandy browns, the polar ice caps in white, and the lush rainforest in dark green.

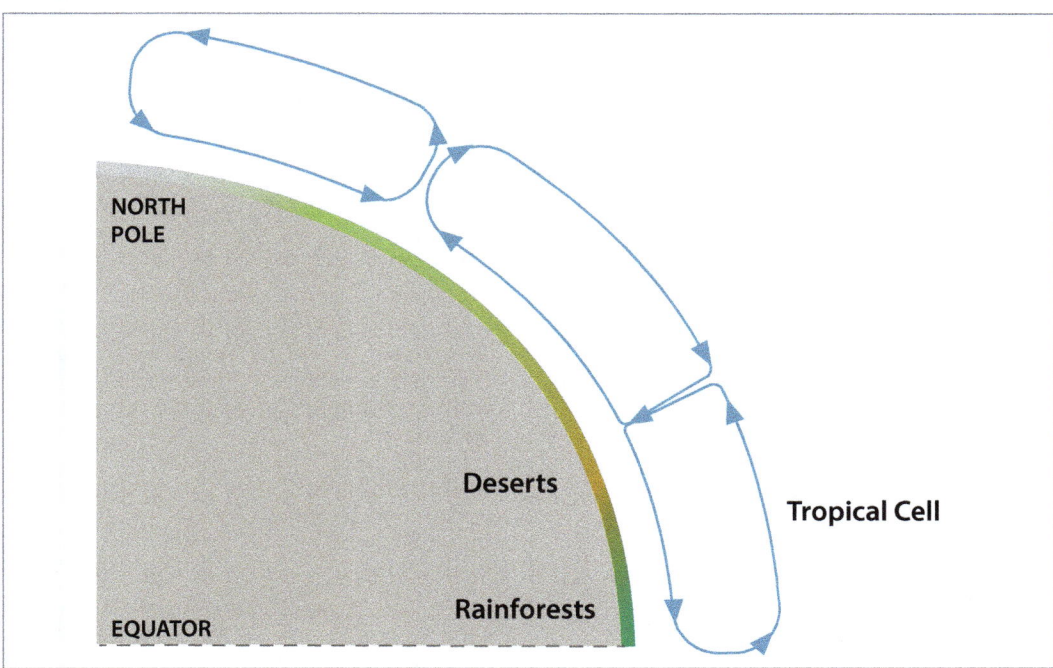

Figure 2.27: **Tricellular model of atmospheric circulation**

Hot, moist air rises at the equator giving a regular pattern, often daily, of convectional rainfall with warm temperatures. The air in this cell moves away from the equator and descends at around 30°N or 30°S, giving a much hotter, drier climate in the areas termed the 'desert belt'. In the desert belt, dry, descending air contains limited water vapour, which gives two subsequent effects. One is the direct limitation of plant photosynthesis due to lack of water as a raw material, but also there is lack of cloud cover and precipitation, leading to higher temperatures than places on the equator that have greater insolation.

Biome name	Biodiversity and Productivity	Climate	Latitude
Tundra	Very little plant life. There are no trees as the soil water is permanently frozen—a permafrost. Lichens survive furthest north. NPP = 10–400 g/m²/year. Diversity is naturally low.	**Polar continental** has the driest climate as the cold conditions lower humidity. There is very low rainfall.	Only found in the highest latitudes, mostly in the Northern Hemisphere. North is ice caps and south is coniferous forests.
Temperate (broadleaf) deciduous forest	Trees grow well, although not as tall as in the tropics. Leaf fall occurs in the autumn as the cold conditions lower the rate of photosynthesis. NPP: 600–2,500 g/m²/year. Diversity is medium.	**Temperate maritime** (coastal) has the wetter conditions needed for tree growth. Mean annual rainfall is 1,000–2,000 mm.	Found in the mid-latitudes, nearer the coasts.
Temperate grasslands, e.g. steppe	These biomes have harsh winters and dry conditions. There is not enough moisture available for tree growth. NPP: around 150–1,500 g/m²/year. Diversity is medium/low; abundance can be high.	**Temperate continental**: cold winter conditions and hot, dry summers are typical. Productive varies considerably increasing with soil water content.	Large areas found mid-latitudes, further inland.
Freshwater lake and stream	These biomes vary in relation to nutrient status and depth. The main producers are either plankton or aquatic plants. NPP: 100–1,500 g/m²/year. Diversity varies with this.	Climate varies considerably, causing large change in nutrient status and productivity. Cold conditions in higher elevations give low production, vice versa for warmer lowland areas.	Various locations.
Deserts	Specialised plants exist in the desert, which can cope with the extreme lack of water, e.g. desert cacti. NPP: 10–250 g/m²/year.	This **arid climate** is found where dry falling air naturally occurs. They are also found inland or in rain shadows. Mean annual rainfall is less than 250mm.	The desert belt is found at around 20–30 degrees north or south of the equator.
Tropical rainforest	A tall, thick canopy at around 40 m, with emergent trees up to 80 m. Animals are abundant but well hidden. NPP: 1,000–3,500 g/m²/year.	**Tropical equatorial**: grows where it rains every day; there is no dry season. Rainfall is greater than transpiration. Mean annual rainfall is over 2,500 mm.	At the equator, from sea level up to 1000 m and mostly between 10 degrees north and south.

Table 2.7: **Summary of continental biomes in terms of structure, biodiversity and productivity**

2. ECOSYSTEMS AND ECOLOGY

Biome name	Biodiversity and Productivity	Nutrient Status
Continental sea	Seas over continental shelves, medium diversity. NPP: 200–600 g/m²/year.	Medium nutrient levels, with seasonal variation. Temperate continental seas have high spring and autumn productivity; nutrient depletes in the summer.
Open ocean	Open oceans are deep seas, with little possibility of nutrients returning from depths. NPP: 2–400 g/m²/year.	Low nutrient levels, open ocean has little primary productivity—the oceans deserts. 92% of marine area.
Deep ocean	Varied biodiversity, with low scientific understanding. Biomes including some biodiversity hotspots. Over 200 m below the surface there is no photosynthesis, but a rain of detritus from above.	Nutrient status is varied, but it is not the key limiting factor. Varied ecosystems such as hydrothermal vents, seamounts and cold-water reefs.
Coral reef	Shallow seas, mostly tropical in distribution. Coral reef has very high biodiversity and productivity. NPP: 500–4,000 g/m²/year.	Low nutrient levels, but coral ecosystems have tight nutrient recycling. Interdependent with ecosystems such as mangrove and sea grass.
Upwelling	Diversity medium, often productive fisheries. NPP: 400–1000 g/m²/year.	High nutrient levels, fed by rising water rich in nitrates and phosphates.

Table 2.8: Summary of marine biomes in terms of structure, biodiversity and productivity

2.4.3 Biome Shifting and the Tricellular Model

Biome shifting is the movement of biome zones due to climatic change. It occurs as the conditions that produce a biome type move geographical location and the biome moves with it. All biomes cannot move as those at high latitude or altitude have no location to move to.

The cells influenced by global warming as they gain more energy for convectional circulation, with the tropical (Hadley) cell expanding, for example. Such changes are an important element of biome shifting with, in this case, the movements of desert belts to higher latitudes. Climate scientists have also predicted that there is a risk of moving to a two cell system with the loss of polar cells in the future.

2.4.4 Succession

Communities go through changes over time as they alter the environment and, in turn, make conditions right for other plants. This process of community change over time is called 'succession'. Positive feedbacks drive change forward.

You need to be able to apply your knowledge of ecosystems theory against case studies. Test your understanding of succession and zonation against case studies.

There are a number of community stages (known as seres or seral stages) that develop in sequence. The final stage is called the climax community. This can be predicted according to the local biome type, depending on edaphic factors (soils) and climate. Therefore, the biome is the climatic climax community of the area.

In a primary succession a barren rock surface is formed following an event such as a volcanic eruption, erosion and deposition activities, or glacial retreat. The surface has, therefore, no developed soil or living organisms present. This rock will first be colonised by pioneer species, such as lichens and mosses. Later a few hardier plants will come in and the soils will begin to develop as dead organic matter falls to the floor and mixes into the soil. This process can be seen following glacial retreat; for example, this sequence was found on exposed boulder clay in Glacier Bay, Alaska.

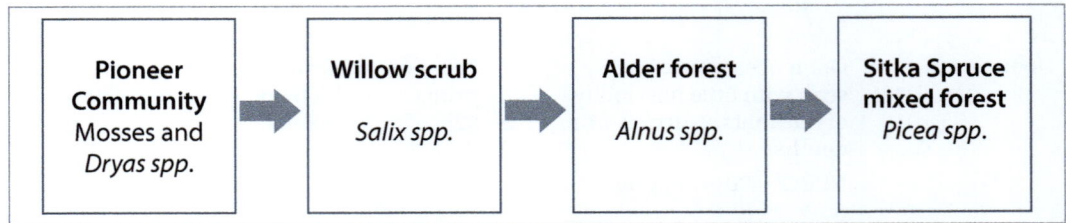

Figure 2.28: **Primary succession following glacial retreat in Alaska**

2.4.5 Human Activity and Succession

Human interference through fire, agriculture, grazing, or habitat destruction can divert succession to a new subclimax, called a 'plagioclimax'. This is an alternative steady state of the system and may persist beyond the original interference or may return through secondary succession to systems similar to the original.

Secondary successions also can result from human interference, for example, when land cleared for agriculture is left unmanaged and goes back through succession, returning to climatic climax communities, but possibly with some loss of original diversity. Abandoned farmland in temperate areas follows a typical pattern as shown by this example from the American East, which occurred as pioneers moved West in the nineteenth century.

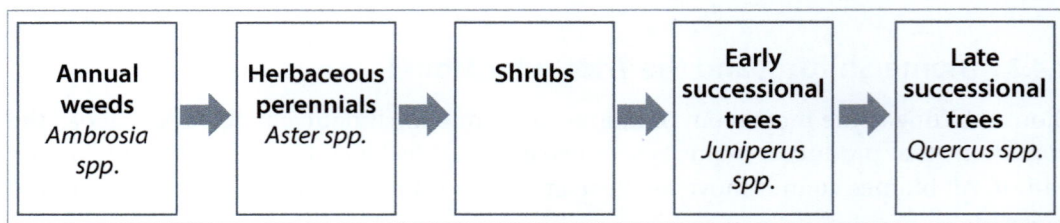

Figure 2.29: **Secondary succession in abandoned farm fields, Eastern USA**

2.4.6 The K- and r- Strategies

Rock and roll

The terms 'r' and 'K' come from population modelling equations, but you can remember 'r' is for 'rock and rollers', the species that live fast and die young. Just remember not to call them that in the exams!

There are a variety of life strategies shown by species and these affect the way their populations develop and their role in succession. The K-strategists produce small numbers of offspring, into which there is a large amount of energy invested from parental care. Population growth is generally S curve style, with the population strongly regulated by density dependent factors. K-strategists are typical of the climax community in succession and are strong intraspecific competitors.

Large numbers of offspring are normally produced by r-strategists. These offspring have very little parental care given and have to survive by themselves immediately. Population growth is more likely to be of a J curve style and the organisms are rapid colonisers whose populations don't stabilise once carrying capacity is reached. r-strategists are often pioneer species in a succession. These life strategies can be placed along a scale as shown in Figure 2.30.

2. ECOSYSTEMS AND ECOLOGY

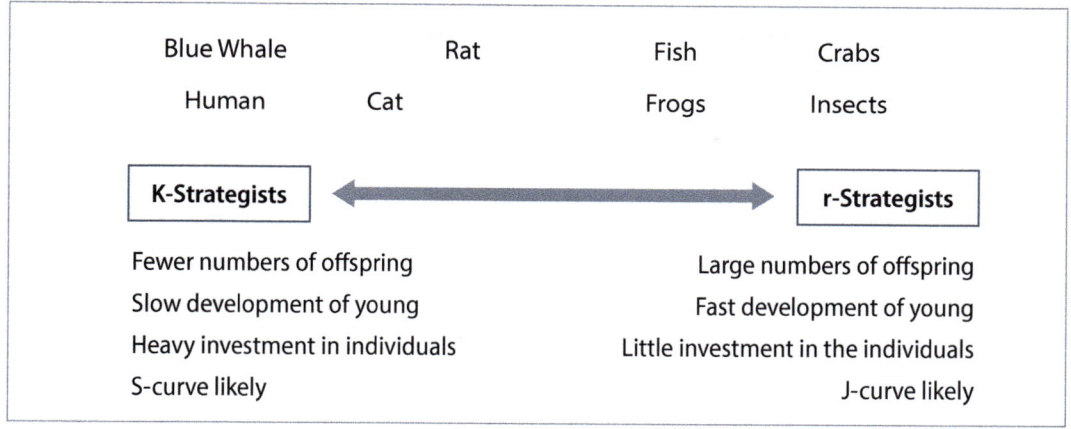

Figure 2.30: **Life strategy scale**

2.4.7 Summary of Changes in Succession

	Soils	(P)hotosynthesis (R)espiration	Biomass	Productivity	Niches	Diversity
Early	Thinner and less fertile	P > R	Low	GP low, NP high	Few, favouring r-selecting species	Low
Mid	Deepening and more fertile	P > R	Medium	GP rising, NP falling	Increasing, still rapid change in species and niches	High
Climax	Mature soil with developed horizons	P = R	High	GP high, NP low or zero	More specialised niches; K-selecting species favoured	High (often slightly lower than mid)

Table 2.9: **Summary of general trends in succession**

When revising this summary table, remember that over time the complexity of the ecosystem increases and that this has corresponding effects to energy and nutrient cycling. In later stages nutrients are cycled more efficiently and energy is channelled along a greater range of pathways. The species strategies change as the opportunist approaches of r-selecting species have an advantage in early stages—K-selecting in later stages.

The capacity of the ecosystem to survive changes from human interference depends on the systems properties such as resilience. In turn these may link to the system diversity.

2.4.8 Zonation

These are changes in communities along an environmental gradient; they occur in many types of ecosystem. Environmental gradients refer to linear changes in a range of abiotic and biotic factors, with communities which occur in distinctive zones.

Species adapted to certain conditions are normally clearly zoned, with some overlap in community types. Zonation examples range from the intertidal shore communities such

> Do not confuse *succession* with *zonation*. Succession is a pattern shown over time; zonation is a pattern shown over space.

as mangroves and rocky shores, beach vegetation, or as altitude changes ascending a mountainside.

Zonations are normally studied using transect methods where samples are positioned along a baseline or transect along the environmental gradient. A line transect method records all species touching the line. Belt transects use quadrats either positioned continuously or, in interrupted belt transects, with regular gaps.

2.4.9 Monitoring Impacts from Oil Spills Using Transects

Baseline surveys and continuous monitoring of rocky shore communities are often carried out using transects. Survey and monitoring has been valuable in assessing the impact of oil spills, monitoring recovery and advising future clean-up operations. The Torrey Canyon oil tanker ran aground in Cornwall in 1967, spilling oil over the Cornish coast and across the channel to Brittany in Northern France. Shores cleaned with toxic dispersants after the spill showed a dramatic growth of green algae, termed 'greening', which indicates a system out of balance. In Brittany signs of a full recovery were not seen for another 10 years.

In 1978 a tanker, the Amoco Cadiz, ran aground spilling oil over the shore again. Despite the bad luck of being hit twice by oil spills, toxic dispersants weren't used in 1978 and, instead, a 6,000 strong team of service men and women physically removed the oil by hand and using portable steam cleaners. The monitoring of the second incident showed an absence of the 'greening' associated with the Cornish spill, and a faster recovery (Southward, 1978).

Practice questions: Biomes, Zonation, and Succession

Explain means to give a detailed account, including reasons or causes. In the context of biomes, the key is to remember how the needs of primary productivity relate to conditions for plant photosynthesis. In turn, these depend mostly on the temperature and precipitation available as dictated by climate.

1. **Describe** and **explain** the relationship between insolation and latitude.

   ....

   ....

   ....

2. **Explain** the significance of this variation in insolation to the rates that plants can grow at different latitudes.

   ....

   ....

   ....

3. **Explain** why temperature and precipitation are key climatic variables influencing biome type.

   ....

   ....

   ....

4. If we remove the influence of altitude and oceans, the kind of biome distribution we would see would look like the simplified model in Figure 2.31 of one hemisphere, only showing change against latitude.

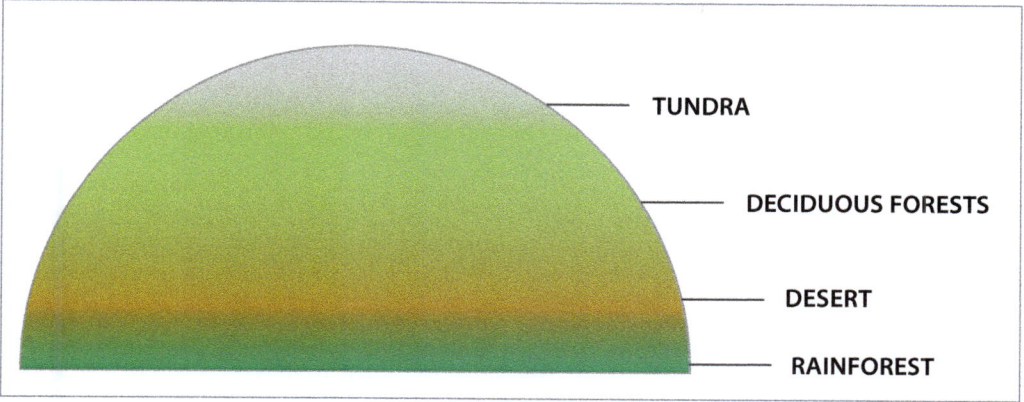

Figure 2.31: **A model of general patterns in biome distribution**

Explain why these four biomes are found at these latitudes.

....

....

....

5. *Give* examples from early, mid, and late succession of named r-strategist and K-strategist species. *Compare* the strategies and role in succession of your examples.

....

....

....

....

6. *Outline* how positive feedbacks move succession forward in a named example.

....

....

....

7. *Evaluate* the use of interrupted belt transects to measure change in the biotic and abiotic factors using an example.

....

....

....

8. Students studying the beach vegetation on Phuket Island used an interrupted belt transect with a regular interval of five metres left between quadrats. Typical zonations looked like the example shown on the next page.

Pioneer Zone	Maritime Grassland Zone	Maritime Scrub Zone	Beach Forest Zone
Sea pea *Canavalia maritima*	Beach sunflower *Wedelia biflora*	Beach fan flower *Scaevolia taccada*	Pacific ironwood *Casuarina equisetifolia*
Goat's foot morning glory *Ipomoea pes-caprae*	Spurge *Euphorbia atoto*	Screw pine *Pandanus odoratissima*	Indian almond *Terminalia catappa*
Sea ⟶			Land

The following year another group of students returned to the beach after the tsunami. The graph shows a change in the diversity on the beach.

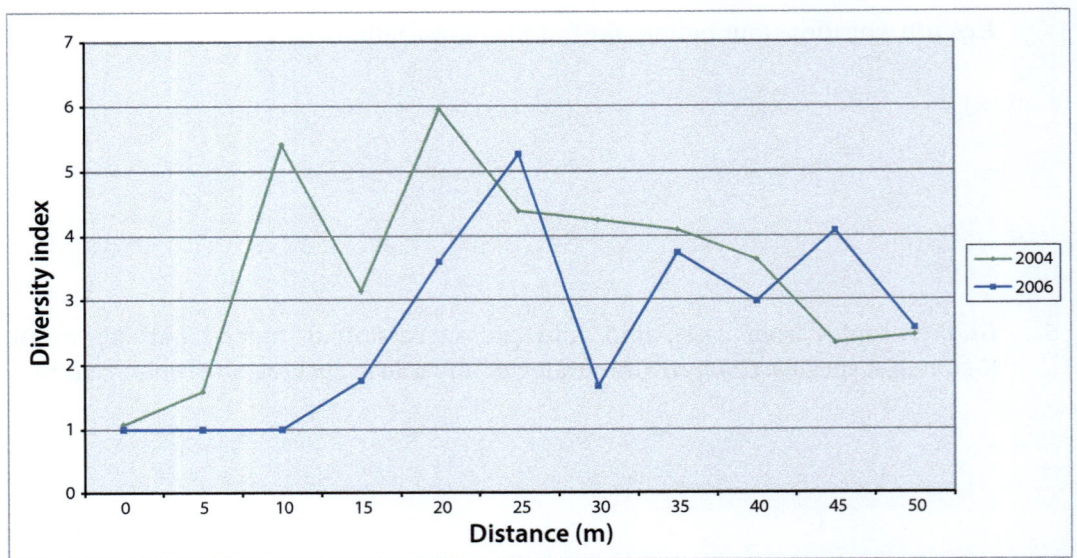

Figure 2.32: **Graph of change in community diversity on the beach in Phuket before and after the tsunami** (the tsunami was on 26/12/2004, after the 2004 data was collected)

(a) *Describe* the pattern of diversity shown before the tsunami hit the beach.

(b) *Compare* the pattern of diversity shown before and after the tsunami.

2. ECOSYSTEMS AND ECOLOGY

(c) The species of plant called goat's foot morning glory grew in a much wider zone after the tsunami. ***Suggest*** a reason for this.

....

....

....

(d) ***State*** what process will eventually restore the old pattern of zonation? ***Draw*** a diagram to show how this will take place.

....

....

....

9. ***Describe*** how a baseline survey of a rocky shore ecosystem could be designed in advance of building an oil refinery and landing pier for oil tankers.

....

....

....

2.5 Investigating Ecosystems

2.5.1 Measuring Abiotic Factors

In general, abiotic factors change through daily cycles, seasonal cycles, short term weather patterns, and tidal changes. In addition, the factors will change with depth or distance from, for example, edges of habitat types or pollution sources. Attention must be given to changes with depth, time or distance as relevant when designing sampling methods. Data logging technology allows continuous data sets to be measured for most abiotic factors. You should be able to describe and evaluate three methods in detail with reference to a named ecosystem.

> Make sure you include the international standard (SI) units in your answers.

Temperature (°C)	Normally measured using thermometers or temperature probes attached to data loggers. Seasonal and diurnal variations are important, as is the influence of aspect.
Wind speed (m/s)	Measured using an anemometer, an instrument with cups that spin in the wind.
Relative humidity (%)	The amount of water vapour measured using a sensor or a hygrometer, as the percentage of possible water vapour at that temperature.
Light intensity (lx)	Measured using a light metre in lux. Seasonal, latitude influence incident radiation.
Light wavelengths (w/m2 for defined wavelengths)	Measured using light probes that can read at particular wavelengths, or colour filters for particular wavelengths. Indicator beads for UV light are available.
Aspect/slope	Measured using a compass to give aspect (direction in which a slope faces), and a clinometer to give slope (gradient).
Elevation (m)	Measured using a hand-held altimeter, GPS, or using map contours.
Soil % organic matter	Assessed by baking in the oven at over 500 degrees to burn off organic matter and which is then given as a percentage of the dry soil mass.
Soil % moisture	Can be measured in the field by using a sensor or assessed by baking in the oven at over 100 degrees to evaporate off the water and given as percentage of original soil mass.
pH	Measured using universal indicator or a pH probe.
Turbidity (m)	Measured in depth using a sechi disc (black and white decorated disc), lowered on a measuring rope until it is no longer visible.
Flow rate (measured in m/s)	Measured using an impeller unit on a stick that measures speed of revolutions, converted to velocity of water or by using a metre stick and a float.
Salinity (% salt)	Measured using a refractometer by placing a droplet of sample water on a lens and allowing light to enter through the water
Dissolved Oxygen (mg/l)	Measured using either an oxygen probe or chemical titration (Winkler's method).
Wave action	Should be measured directly by measuring the size and frequency of the waves or by using biological scales such as Ballantine's exposure scale.

Table 2.10: **Abiotic measures used in ecosystem investigation**

2.5.2 Species Identification

When making keys, start with a feature that divides the group into two similar sized groups first. Then focus on the smaller differences within each

This is usually done with a published identification key. The key asks a question and the answer determines what step to go to next, either the name of the species or another question.

For example, Figure 2.33 is a simple key to distinguish insects, spiders, and millipedes.

Question	Answer
1. Does the animal have more than three pairs of legs?	Yes: Go to next question No: Insects
2. Does the animal have four pairs of legs?	Yes: Spiders No: Go to next question
3. Does the animal have many pairs of legs?	Yes: Millipedes

Figure 2.33: **Summary of methods for measuring abiotic factors**

You need to be able to make your own key for up to eight different organisms.

2.5.3 Estimating the Abundance of Organisms

Having identified which species are present some idea of abundance is often needed. This is often done within a defined sample area termed a 'quadrat'. The quadrat size depends on the size of the organisms in the habitat and commonly accepted standards. On grasslands, a common size of quadrat is 50 × 50 cm. For plants, algae, and some colonial animals like coral, it is usual to estimate the abundance of a species by either a percentage cover estimation or calculation of percentage frequency.

Figure 2.34: **A point quadrat**

Percentage cover is a two-dimensional estimate of the spread of the organism within a simple frame quadrat as a percentage. This method is fast but highly subjective. However, if the same user collects data in two or more areas, then it may be useful for comparisons.

$$\frac{\text{No. of occurrences in the quadrats}}{\text{No. of quadrats taken}} \times 100\% = \text{Percentage frequency}$$

Percentage frequency can also be collected using quadrats. Using this method, the number of occurrences is given as a percentage of the total possible number of occurrences.

This is an objective measure, but it does take time. It becomes a good measure of abundance if a large number of small quadrats are used, ideally over 100 very small quadrats. In the point quadrat design shown, a t-bar frame holds ten needles—each of them a single quadrat sample.

2.5.4 Estimating Biomass

Biomass is the mass of living material in the ecosystem. It is measured as dry mass where possible. Look back to the exercise on page 59 to see how laboratory measurements of invertebrate dry biomass were used to estimate biomass in trophic levels.

For the estimation of biomass of plants within a quadrat, they are sampled destructively. The plants are dug up, dried, and weighed. This dry weight per unit area can then be used to estimate average biomass per quadrat, or the total biomass in the habitat. This method is fairly straightforward in habitats such as grasslands and can be used to estimate productivity.

In woodlands, measurements of tree size are often used by measuring girth at chest height and calculating the volume of the tree. This can be converted to mass using secondary data on the density of wood.

Practice questions: Investigating Ecosystems 1

1. *State* the name of an ecosystem that you have studied and *list* the most significant abiotic factors for that ecosystem.

2. For three of the abiotic factors given *describe* the methods you would use to measure them with variations against depth, time, or distance.

3. *Evaluate* how well these methods you have described would represent the actual variations in the ecosystem.

4. **Construct** a key to help identify the following eight species of tree from the leaf outlines given.

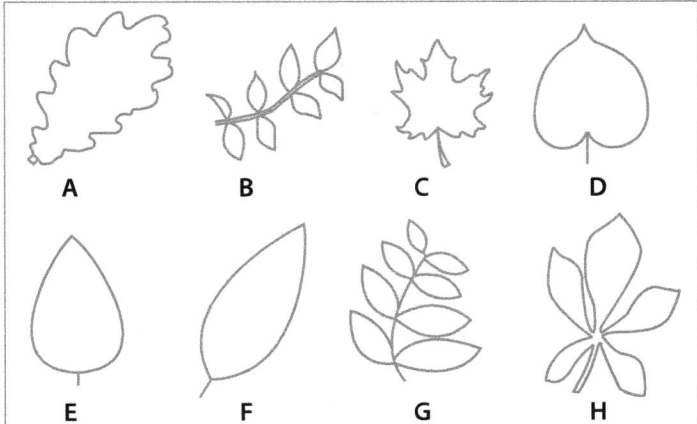

Figure 2.35: **Leaves from eight species of UK tree**

Use the following terms to help write the key:

- 'Simple leaf' for a single leaf structure
- 'Compound leaves' divided into many leaflets
- 'Lobed' for large indentations
- 'Toothed' for small indentations
- 'Palmate compound' for whorled leaflets arising from a single point

....

....

....

....

5. A 10 × 10 m area (100 m²) is marked out in a 5 hectare (ha) grass field and 1 × 1 m quadrats are positioned using coordinates generated by random numbers. 25 quadrats were sampled destructively, with the grass removed, dried and weighed.

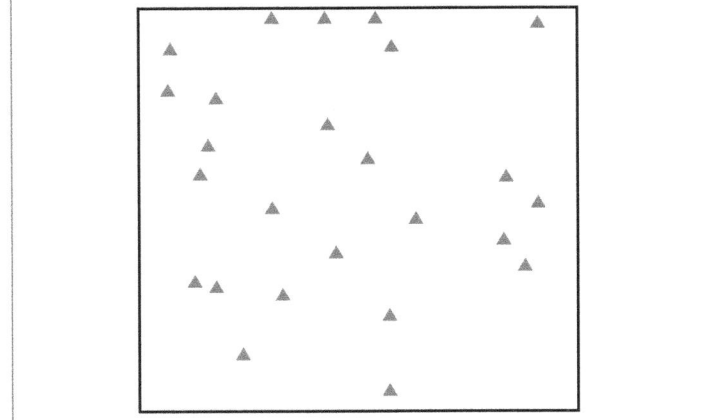

Figure 2.36: **Random sampling in grass field**

The average dry mass of grass in all quadrats was 689 g.

This method was repeated in another part of the field after one year of no cutting and no grazing. The average dry mass of the grass was now found to be 1,112 g.

(a) **Determine** the percentage of the 10 × 10 m area which was sampled.

....

....

....

(b) **Determine** the standing crop biomass for the 10 × 10 m area in the first year.

....

....

....

(c) **Calculate** the growth rate per m², in grams over a full year.

....

....

....

(d) **Calculate** the daily growth rate.

....

....

....

Note 1 ha is 10,000 m².

(e) **Determine** how much hay (dry grass) could be harvested from the whole field in a year.

....

....

....

(f) **Evaluate** this method.

....

....

....

6. ***Estimate*** the percentage cover of leaves in the quadrat shown in Figure 2.37.

   ....

   ....

   ....

   ....

   ....

   ....

Figure 2.37: **Leaves in a frame quadrat**

7. ***Determine*** the percentage frequency of leaves found in the 100 smaller squares of the gridded quadrat.

   ....

   ....

   ....

   ....

   ....

Figure 2.38: **Leaves in a gridded quadrat**

8. ***Determine*** the percentage frequency of all the leaves found under the points of all 121 intersections in Figure 2.38. Collect the results from others working independently of you if possible. Compare the ranges and averages.

   ....

   ....

9. ***Evaluate*** these two techniques.

   ....

   ....

   ....

   ....

ENVIRONMENTAL SYSTEMS AND SOCIETIES SL

2.5.5 Direct Methods of Estimating Abundance in Animals

Animals that don't move quickly, such as rocky shore limpets or grassland snails, can be counted in quadrats giving a direct measure of population density. This is only suitable for species that don't run away. A variety of direct sampling techniques can be used to collect invertebrates using nets and traps. They need standardising.

> **Standardising:**
> For both direct and indirect methods, samples must be standardised. Quadrats sample standard area. Other techniques need standard effort, such as the amount of time or the number of kicks.

Here are some:

- Freshwater nets for lake and stream invertebrates (invertebrates kicked into the net)
- Sweep nets for grassland and scrub (Invertebrates swept into the net)
- Pit trapping and baited traps for terrestrial invertebrates (invertebrates trapped or attracted by bait)
- Beating trays for invertebrates in trees (invertebrates shaken from the branches onto white cloth held on a frame).

2.5.6 Indirect Methods of Estimating Abundance in Animals (Lincoln Index)

There is an indirect method of establishing population size called the mark, release, and recapture technique. Firstly, a sample is taken from a population and the individuals marked, then the marked animals are released back into the habitat to mix with the original population. A second sample is then taken from the population and a count made of the animals that are marked and the total of all sampled. N is therefore calculated from the Lincoln index.

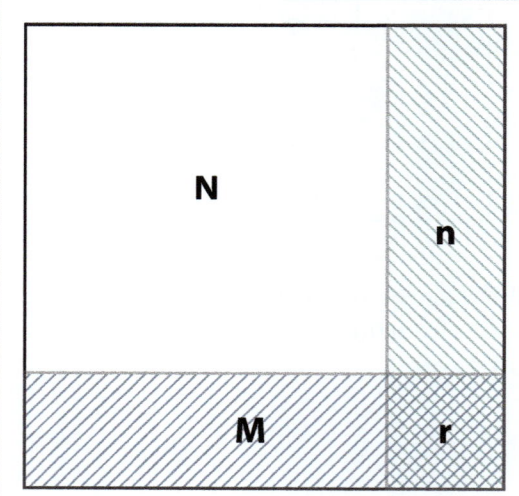

$$N = \frac{M \times n}{r}$$

N is the total number in the population (the whole square, including M, n and r)

M is the total number caught in the first sample (includes r)

n is the total number caught in the second sample (includes r)

r is the total number recaptured (marked) in the second sample

Figure 2.39: **Diagram showing the Lincoln index**

The method assumes that the proportion on N:M should be the same as the n:r.

When evaluating this technique, you should consider the following assumptions of the index:

- Animals are as likely to be trapped in both sample periods.*
- The marks do not harm the animals.
- There is no change in the population during sampling.
- Animals mix freely back into the population after marking.
- Sample sizes are of similar effort.
- The second sample has at least 10% recapture.

*Both trap-happy and trap-shy animals occur. The former like the food and comfort in the trap; the latter won't go near the trap again.

2.5.7 Diversity and the Simpson's Index

Diversity is a useful concept for analysing variety in systems. When ecologists calculate the diversity of species in a community, they are looking at the two measurable components of the community. Firstly, there is species richness, **the number of species**, which can be found by identifying species and creating a species list. Secondly, they look at the **relative proportions** of species in the community.

Diversity indexes such as the Simpson's index can be used to give a value to diversity. Here is the formula to calculate Simpson's as D:

$$D = \frac{N(N-1)}{\sum n(n-1)}$$

N is the total number of organisms in the community.

n is the total number of organisms of an individual species.

$\sum$ is sigma, the sum of.

To calculate the bottom of the equation, draw up a table like this example of data from a single sweep net sample.

Species	n	$n-1$	$n(n-1)$
Red spiders	8	7	56
Grasshoppers	12	11	132
Bush crickets	8	7	56
Click beetle	1	0	0
$N =$	29	$\sum n(n-1)$	244

$n(n-1) = (29 \times 28) = 812$ $D = 812/244$ $D = 3.328$

Formulas:
You do not need to memorise the formulas for the Lincoln or Simpson's indexes. However, you should be able to use them in an exam, including knowing what the symbols stand for.

2.5.8 Interpreting the Simpson's Index

Simply put, the lower the value of D, the less species diversity there is. The lowest value possible is 1; the highest value possible depends on the number of species in the sample and on their proportions. The total number of organisms has little influence on the value of D.

This can be a useful tool for comparing similar habitats to see if there is a significant difference in diversity. This needs to be done carefully; some communities have a naturally low diversity, for example, Arctic ecosystems. For further discussion see 1.3.10 and 3.1.3.

If similar communities are different in diversity it may be caused by human impact, see, for example, the impact of pollution on freshwater invertebrates (4.4.5). The stage of succession also influences diversity with typical patterns showing a general increase, which may fall slightly at climax (see table on succession: 2.4.7)

Diversity indices
Be careful as there are several diversity indexes, including others with Simpson's name. They have different ranges and values for diversity. **Best to avoid looking online!**

ENVIRONMENTAL SYSTEMS AND SOCIETIES SL

Practice questions: Investigating Ecosystems 2

1. A sample of 131 lemmings are trapped in baited mammal traps and marked with permanent pen on the inside of their ears. The trapping period is a full 48 hours. The same number of traps are positioned in the same places one week later. 143 lemmings are trapped, of which 27 are marked.

   (a) Use the Lincoln index to *estimate* the number of lemmings in the trapping area.

   (b) *Evaluate* the method.

2. Two farm ponds are shown below, with four species of fish found: red headed, spotted, blue striped, and rainbow.

   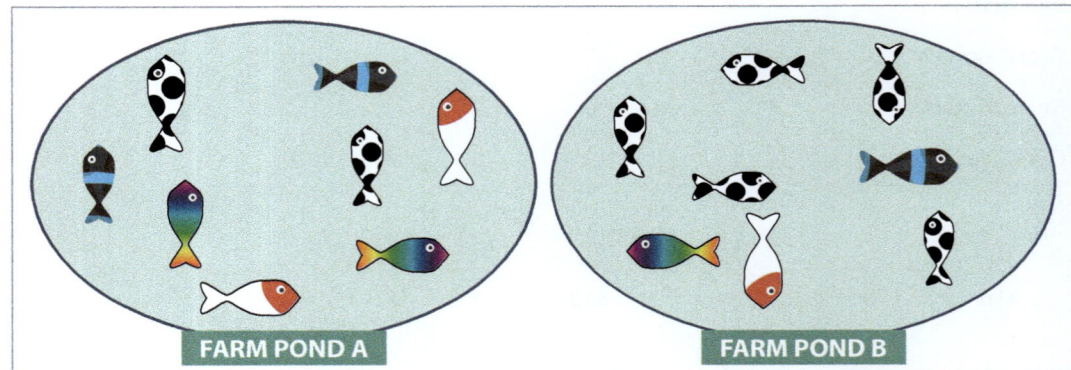

   Figure 2.40: **Farm ponds A and B**

   (a) *State* which pond appears to be more diverse.

   (b) Test your answer to (a) by *calculating* the Simpson's index for each pond.

   (c) One pond has lower oxygen due to organic pollution. *State* which pond you think it is.

Chapter 3: Biodiversity and Conservation

3.1 Introduction to Biodiversity

Biological diversity, or simply biodiversity, is a term used by conservationists and politicians to describe total variety within ecosystems. It is defined as follows:

The amount of biological diversity or living diversity per unit area, it includes the concepts of species diversity, habitat diversity, and genetic diversity.

3.1.1 Components of Biodiversity

Term	Definition
Species diversity	The variety of species per unit area. This includes both the number of species and their relative abundance in the community.
Habitat diversity	The range of different habitats or number of ecological niches per unit area in an ecosystem, community, or biome.
Genetic diversity	The range of genetic material present in a gene pool or population of a species.

Table 3.1: **Definition of key terms used in defining biodiversity**

3.1.2 Richness or Diversity?

The number of species present—or species richness—is often used to indicate general patterns of biodiversity and can be used to compare habitats, biomes, countries, or indicate global biodiversity. It is also a useful measure to investigate damage to biodiversity over time. Remember that this is not species diversity, and, additionally, it has no element of habitat or genetic diversity, so it only indicates biodiversity. There is no simple way to quantify biodiversity.

3.1.3 Explaining Biodiversity

Comparing different ecosystems or biomes in terms of biodiversity allows us to consider the reasons for patterns in diversity that are present in natural communities and allow us to understand the system stability (1.3.10). Studying biodiversity is important in conservation as it allows us to both understand and assess the conservation value of a community. All else being equal in a similar community then lower diversity tends to indicate pollution, physical disturbance, eutrophication, or invasion by a highly competitive species. A lowering of biodiversity tends to mean that one or a small number of species are dominating the system.

What is known as the conventional wisdom in ecology can be stated as, 'a complex ecosystem with its variety of nutrient and energy pathways provides stability', as in the IBO course guide. Or more simply put by Charles Elton, 'Complexity begets stability' (Elton, 1958). This is shown, for example, in complex ecosystems such as tropical rainforests. Rainforest ecosystems appear to be stable systems, for example, resistant to invasion by foreign species. However, if they are disturbed stability is low, as the system takes a long time to return to equilibrium. Polar ecosystems are less complex and yet there is considerable stability in these systems. The pattern is not simple.

One answer to the apparent relationship is that it is stability that leads to diversity. This is shown, for example, in the glacial refuges of the South American rainforest. Here conditions remained as tropical rainforest even during the ice ages. Consequently, there has been a long time for diversity to develop in these areas and the highest levels of recorded diversity are found there. Other theories suggest diversity requires regular disturbance, to avoid one species dominating. This can be found, for example, in the high plant diversity of heavily grazed sheep fields.

Practice questions: Biodiversity

1. Complete the following table of definitions:

Term	Definition
Biodiversity	
Species diversity	
Habitat diversity	
Genetic diversity	

2. ***Discuss*** the difficulties of accurately estimating the number of species on the planet. ***Explain*** why these estimations are important to the conservation of biodiversity.

....

....

....

....

....

3. BIODIVERSITY AND CONSERVATION

3. A coral reef is surveyed by divers and transects are carried out to record species richness and to calculate a diversity index. The divers also carried out a reef health index based on physical features relating to structure such as coral coverage and fish biomass.

Results of Reef Survey

Site	Richness	Diversity	Reef Health Index
1	7	1.25	1.0
2	9	1.50	1.3
3	17	1.75	2.0
4	15	1.45	3.0
5	13	1.85	3.2

Reef Health Index: Key	
1	Critical
2	Poor
3	Fair
4	Good
5	Very Good

(a) For sites 4 and 5 *describe* how is it possible that the calculated diversity in 5 is higher, but richness in 4 is higher?

....

....

....

....

(b) A scientist claims that the reef health index is an effective method for designating conservation areas to protect biodiversity. Use the data and graph provided to show why this may not be true. *Suggest* a reason why this may be the case.

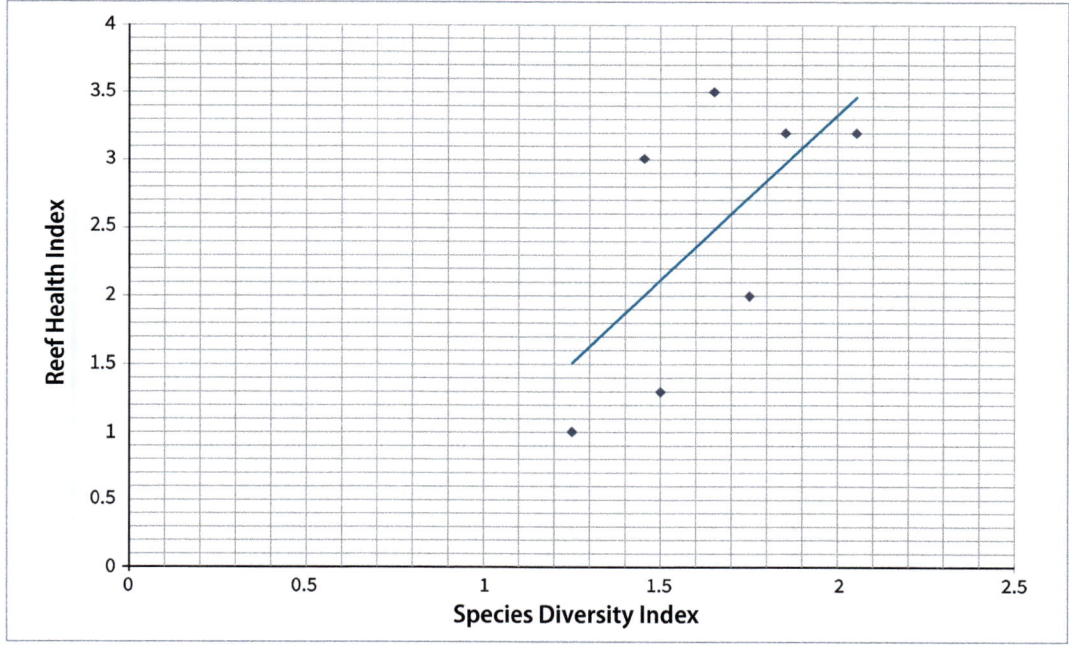

Figure 3.1: **Scatter graph showing correlation of a species diversity index and reef health index**
Source: Data from Díaz-Pérez, Rodríguez-Zaragoza, Ortiz M, JD, & Ríos-Jara, 2016

(c) Species identification work and the calculation of diversity requires specialist training, whereas the reef health index approach can be learned quickly. Assessing richness is quicker than diversity. *Evaluate* appropriate survey techniques as follows:

(i) Which would you recommend for surveying physical damage to coral reef around tourist areas?

....

....

(ii) Which would you recommend for recording long term change to reef due to coral bleaching caused by warming oceans?

....

....

(iii) Which would you recommend for ecological investigation to better understand the structure of the community?

....

....

....

3.2 Origins of Biodiversity

3.2.1 Evolution

Evolution is a gradual change in the genetic character of populations over many generations, a process which has taken 3 billion years to develop today's biological diversity. Evolution can lead to the development of new species; this is called speciation. Speciation can occur due to isolation of individuals in a new situation. Environmental change gives new challenges to the species, those that are suited survive, those that are not will not. Through the mechanism of natural selection new species may arise over long periods of time.

3.2.2 The Mechanism of Natural Selection

Natural selection is an evolutionary driving force, sometimes called 'survival of the fittest'. In this context the meaning of fittest is understood as 'the best suited to the niche'.

Adaptation

While it is true to say organisms are adapted to their niche (or they are dead), be careful with this term.

Misuse of the word can give the impression that the organism is actively seeking a change or that the change is occurring very quickly.

3. BIODIVERSITY AND CONSERVATION

Natural selection is one of the main methods in which speciation occurs through the following steps:

- Within a population of one species there is genetic diversity, which is called variation.
- Due to natural variation some individuals will be fitter than others.
- Fitter individuals have an advantage and will reproduce more successfully.
- The offspring of fitter individuals may inherit the genes that give the advantage.

Due to the process of natural selection over generations, there is a movement towards fitness in the whole population, the environment (niche conditions) constantly selecting for the individuals that are best suited to that niche. Negative changes lead to extinction.

3.2.3 Isolation

Isolation of sections of a population can happen due to geographical processes forming barriers such as mountain building, changes in rivers, sea level change, or plate movements. Sometimes natural selection for different environmental conditions leads to new species separated by a physical barrier.

Similar isolation can occur when a group of pioneers from the population colonises a new niche. Natural selection can lead to changes that create biological barriers to reproduction, which can be behavioural or physical. Isolation can also occur within populations due to genetic changes that take place leading to physical or behavioural barriers to reproduction.

All forms of isolation can lead to speciation, that is, when the two isolated sections of the population are no longer able to interbreed and produce fertile offspring. The gene pools have become permanently separated, meaning there is no longer any active mixing of genes between the two species.

Learn examples

Longer answers are much better when illustrated by specific examples. Populations of Galapagos finches (Darwin's finches) are often isolated. For example, the Large and Common cactus finches are found on different islands

3.2.4 The Influence of Plate Activity on Evolution and Biodiversity

The surface of the earth is divided into eight major crustal plates and several minor ones, which have moved throughout geological time. This has led to the creation of both land bridges and physical barriers with evolutionary consequences. The distribution of continents has also affected oceanic and atmospheric circulation patterns, causing climatic variations and variation in food supply.

Plate movements influence evolution and biodiversity over geological time, Table 3.2 shows a summary. Here the number of families and genera (plural of genus) is used to indicate biodiversity.

3.2.5 Biodiversity Rising

Key points in Table 3.2:

- There is a general rise of diversity over time.
- Speciation rates must be higher than extinction rates *overall* (not at times of diversity loss).
- Extinction events are followed by periods of adaptive radiation.
- The number of continents positively correlates with diversity.

Species diversity rises despite five major extinction events. Extinctions lead to evolutionary recovery (radiation of new species).

Age (Millions of Years: mya)	Diagram of Plate Distribution	Description	Evolution	Biodiversity Level (Number of Families)
Over 600		One super continent	Soft bodied multicellular organisms present	Under 100
500–599		Super continent divides into four continents	Exoskeletons develop in many groups. Radiation of invertebrates	Expansion to 150
450 mya mass extinction (27% of all families, 57% of all genera)				
400–499		Rejoining of two continents	First land animals and plants	Expansion to 280
360 mya mass extinction (19% of all families, 50% of all genera, 70% of all species)				
300–399		Continents continue to rejoin	Amphibians and reptiles	240
251 mya mass extinction (57% of all families, 83% of all genera, 85% of all species)				
250–299		Formation of a supercontinent	Age of the dinosaurs	220
205 mya mass extinction (23% of all families, 48% of all genera)				
200–249		Formation of Laurasia (N) and Gondwanaland (S)	Radiation of flowering plants and insects	160
100–199		Fragmentation into sections that develop into the modern continents	Age of mammals	300
65 mya mass extinction (17% of all families, 50% of all genera)				

Table 3.2: **The influence of plate activity on evolution and biodiversity**

Continental division provides barriers for land-based species and land bridges present barriers for marine species. These lead to speciation due to isolation of the gene pools. Due to plate movements, environmental conditions change continuously, leading to varying effects from natural selection.

Species diversity is also associated with the climate and growing conditions. Fragmented continents have a more oceanic climate, leading to better growing conditions on land. Supercontinents produce continental climate types, with harsher conditions (cold winters, hot dry summers) in the continental centre. Variations in latitude of the land surface influence global climate patterns.

However, when drawing conclusions, remember that knowledge of the fossil record is patchy (incomplete) depending on the environment of formation and on the research interest in different locations.

3.2.6 Mass Extinctions

There are five major mass extinction events in the geological past and we are currently experiencing a sixth. These events are different to the background extinction rate explained in Table 3.2, as they are catastrophic events with very high species extinctions.

There are many debates on the ultimate cause of the different mass extinctions, including plate collisions, meteorite impact, super-volcanic eruptions, or lava flows. They are often also associated with sudden changes in atmospheric composition, sea level rise, anoxic conditions (low oxygen), and climatic change. Primary productivity falls with subsequent impacts on the food chain, whilst other planetary life support systems fail.

The largest extinction event was the permo-triassic extinction 250 mya, or the 'Great Dying'. Some entire taxonomic groups were wiped out, such as trilobites, and marine life in general experienced 96% extinction levels. Biodiversity was so badly damaged that it took over ten million years to recover to the previous levels.

3.3 Threats to Biodiversity

3.3.1 Global Biodiversity Assessment

Globally the total list of recorded species in 2010 was over 1.7 million, compared with around 1.4 million in 1990. The list is growing by around 15,000 species a year, not due only to new discoveries, but also as research on existing named species shows they consist of more than one species (one species therefore becomes two).

Estimates for species that are currently nameless are much greater; estimates range between 20 million and 100 million. Research on rainforest beetles, by insecticide fogging of enclosed trees, has suggested that in this group alone there may be 20 million species on Earth (Erwin, 1988). Such estimates are based on mathematical models projecting from observed data.

In other words, the percentage of known over unknown species is at best just over 8%, at worst under 2%. Many areas are unrecorded due to difficulty of access and lack of specialists for taxonomic groups found, for example, rainforest canopies, microbes, or soil invertebrate communities.

Group	No. of species
Viruses	5000
Bacteria	9000
Fungi	100,000
Protoctista*	60,000
Plants	300,000
Invertebrates	1,200,000
Vertebrates	60,000
Total	1,734,000

Table 3.3: **Approximate numbers of named species**
a group that includes protozoans and algae
Source: Includes data from Wilson, 2001

Another area that is under-explored is the abyssal plain. This area is the bed of the deep ocean, over 3km down and covers over half the surface area of the Earth. 80% of species retrieved in each visit are new to science. As well as difficulty in accessing some areas, there can be difficulty in identifying many taxonomic groups and a shortage of specialists.

3.3.2 Natural Threats to Biodiversity

Large numbers of species have been wiped out by natural catastrophes in mass extinctions. There is also a background extinction rate which occurs due to both natural events and replacement evolution (when older ancestral species are replaced slowly by a newer version). The fossil record shows that most species that have evolved have later become extinct. Natural events that lead to species extinction are diverse and are similar but may be more localised than those leading to mass extinctions. Such factors include meteorite impact, volcanic eruptions, droughts, ice ages, climate change, ocean acidification and anoxia, sea level rise, and genetic inferiority.

3.3.3 Background Extinction Rates

The fossil record shows that on average a species has a predicted lifespan of 1–10 million years. For some groups the lifetime seems much shorter than others: mammal species in the last 65 million years, for example, live on average of 1–2 million years, whereas invertebrate species may last over 10 million years.

Extinction rates have been calculated for a period stretching over the last 600 million years. Background extinction has been calculated at 9% in every million years, or around one species every five years. Remember that rates are calculated from an incomplete fossil record (Raup, 1988).

3.3.4 Present Day Extinction Rates

The current rate of extinction, though, is at least 100 times the background rate and possibly as high as 10,000 times the background rate. Current estimates range from 30,000 to 60,000 species a year. Current predictions suggest that, if environmentally destructive human behaviour does not change, we will have lost fifty percent of the planet's species by 2100. However, as we don't currently know the total number of species on the planet, it is impossible to be certain about the rate of extinction of species.

The human impact started at least 100,000 years ago when hunter-gatherer communities may have caused extinctions of large animal prey in North America and Australia. Human impact intensified at the birth of agriculture 10,000 years ago. Records show that most extinctions since 1500AD have been caused by humans (May & Lawton, 1995).

3.3.5 Human Threats to Biodiversity

The reasons that humans have had such a large impact on extinction can be remembered as 'A HIPPO':

- **A**griculture
- **H**abitat destruction and degradation
- **I**nvasive non-native species
- **P**ollution
- **P**opulation
- **O**verharvesting

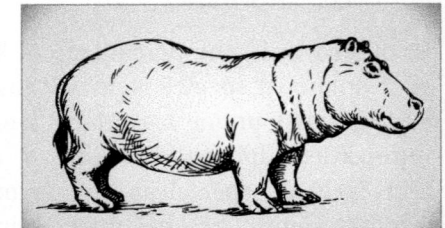

Figure 3.2: **A Hippo**

3. BIODIVERSITY AND CONSERVATION

Agriculture is often a direct cause of habitat destruction and can also lead to loss of genetic diversity through selective breeding. Diversity associated with older styles of agriculture, such as the fallow field, are lost when replaced with intensive monocultures.

Habitat destruction and degradation is caused through a variety of activities such as commercial logging of native forests for timber, road building, settlement, drainage of land for agriculture, slash and burn agriculture, urbanization, and mining.

Invasive non-native species, sometimes referred to as 'aliens'; they are species that are not naturally found in a habitat that have competitive advantage over existing ones or fill an empty niche. Due to the global mobility of people and goods, they have spread between countries all over the globe.

Pollution has a range of impacts on different groups; for example, acid precipitation is a particular threat to amphibians. Climate change is a consequence of pollution that has become a significant impact and has been associated with some recent extinction.

Population growth is bringing more people into contact with the remaining conservation areas, particularly in LEDCs; they may be a direct threat to the species through hunting and gathering activities.

Overharvesting of species beyond sustainable yields can drive reduced populations in small fragments of habitat into extinction.

Often it is more than one factor that leads to extinction, an endangered species reduced to small numbers can be wiped out be a single catastrophic event.

Species	Last sighting	Reasons	Location
Scimitar-horned oryx (*Oryx dammah*)	1996	Habitat loss; Hunting	Chad
Golden toad (*Bufo periglenes*)	1989	Disease; Pollution; Global warming	Costa Rica
Schomburgk's deer (*Rucervus schomburgki*)	1938	Habitat destruction; Hunting; Agriculture	Thailand

Table 3.4: **Multiple causes of extinction in three species**

3.3.6 Red List of Threatened Species

The International Union for the Conservation of Nature (IUCN, also known as the World Conservation Union) has collected data on endangered species for the last four decades. The IUCN screens species against several categories to determine conservation status (as given for each example in the table previously). The IUCN uses several key factors when determining conservation status, some of the key factors are summarised in Table 3.5.

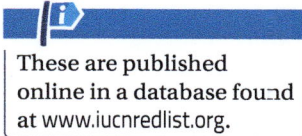
These are published online in a database found at www.iucnredlist.org.

Conservation status is determined by both the changes the species faces in terms of external threats to habitat quality and how the species responds to these changes. In terms of the response of the species, the IUCN considers a wide variety of factors in relation to niche and reproduction, including reproductive potential, trophic level, behaviour and degree of specialisation.

Criteria	Critically Endangered	Endangered	Vulnerable
Population size and reduction over ten years	≥ 90%	≥ 70%	≥ 50%
Geographic range	≤ 100 km²	≤ 5,000 km²	≤ 20,000 km²
Area of occupancy	≤ 10 km²	≤ 500 km²	≤ 2,000 km²
Number of fragments	= 1	≤ 5	≤ 10
Numbers of mature individuals	< 250	< 2,500	< 10,000
Probability of extinction	≥ 50% in ten years	≥ 20% in 20 years	≥ 10% in 100 years

Table 3.5: **Determining species conservation status**

3.3.7 Case Histories of Three Species

You need to be able to describe the case histories of three species: one extinct, one critically endangered, and one whose status has been improved by intervention. Three histories are outlined in Table 3.6 on page 91.

3.3.8 Tropical Biomes

Tropical biomes have particularly high biodiversity, as is exemplified by rainforests and coral reefs. Other tropical systems also have high biodiversity (for example, tropical seasonal forests) and often demonstrate interdependence between ecosystems (such as mangroves and coral reefs).

Tropical rainforests are seen as particularly vulnerable ecosystems, due to deforestation rates and slow speed of regeneration. Without assistance, deforested areas of rainforest may not even regenerate to rainforest biome type if the impact on local climate has been particularly great. The biodiversity per unit area and the total biodiversity in rainforests is great. Studies in Indonesian rainforests (Kalimantan, Borneo) record 470 species of tree per ha whereas in a typical temperate forest this number is under 10 per ha. Other systems have high biodiversity per unit area but do not cover so much space.

Coral reefs have high biodiversity and are vulnerable to human impact, but they do not cover the large land areas of rainforest. Rainforests still cover over 5% of the planet's land surface (reduced from over 10%); this large area and high diversity per unit area means they are home to over 50% of the planet's species, many of which are endemic (Connell, 1978). For further details of an example of a tropical forest, see the case study of Northern Thai hill forest in chapter 3.3.11 on page 92.

Endemic species
Species that evolved in an area and are restricted to that location.

3.3.9 Damage to Ecological Services

Forests regularly regenerate following disturbance on a small scale due to natural tree death and gap formation. Clear felling of forests presents great problems for regeneration, though, as the regeneration of trees requires elements of biodiversity that may be no longer present. For example, fig trees need fig wasps for pollination and birds for dispersal. Many trees need fungi in the soil to help seeds germinate and to grow effectively.

With the canopy removed, the thin soils can quickly wash away and become exposed; drying soils may be prone to fire. The change in the physical conditions means that trees may not grow as tall as they would in an undisturbed forest, leading to a lower canopy height and a lower diversity. Overall, the regeneration of a clear-felled rainforest could take hundreds of years. Acceleration of regeneration is possible when tree species are carefully selected and researched before planting (Elliott, 2005).

Species	Summary of Causes	Socio-political and Economic Pressure	Consequences to Biodiversity
Extinct (1938) Schomburgk's deer (*Rucervus schomburgki*)	Hunting and habitat destruction of swamp forests around Bangkok, Thailand. An expedition to find wild deer for captive breeding failed, due to a translation error of the name they were sent to the wrong part of the country. The last deer held in captivity was mistakenly clubbed to death.	The deer was hunted by gangs, surrounding the herds on islands in the swamp. The animals were killed with clubs and spears. The deer also possessed beautiful antlers, thus putting a price on its head. The deer's habitats were drained, cut down, built on, or turned into paddy fields. Drainage of swamps to make productive rice paddy marked the beginning of the rice export business.	A symbol of a habitat that is lost now to the suburbs of Bangkok. The extinction of this species was seen as a catalyst for the need of conservation of wild flora and fauna, thus helping the case for the establishment of the Thai Royal Forest Department in 1961. (Lekagul & McNeely, 1977)
Critically endangered South China tiger (*Panthera tigris spp. amoyensis*)	Persecution, habitat degradation and fragmentation and low prey density. Critically endangered status as the tiger has not been seen in the wild since the 1970s	A healthy population of 4,000 existed in the 1950s when a programme of eradication was started; to remove the tiger as pests along with their habitat. This was part of Mao Zedong's government policy to convert forest to farmland to promote economic development. Surveys suggest that this species may be extinct in the wild, conclusively it has been extirpated (locally extinct) in 12 of 16 countries that it used to live in.	Further survey work is needed to establish clearly if it is in the remaining four countries. Captive populations of 57 individuals from 6 parents show signs of reduced genetic diversity. Reintroductions into the wild are planned in China. This sub-species is one of only six groups of tiger left from the original 9. All are endangered or critically endangered. The South China tiger is considered to be the ancestral tiger and contains distinct genes.
Previously extinct in the wild Arabian oryx (*Oryx leucoryx*)	Hunting. The last animal in the wild was killed in 1972.	Having once lived throughout the Arabian peninsula this animal was widely hunted when rifles arrived in the mid 19th century, with four wheel drive vehicles intensifying pressure in the mid 20th century. The Fauna Preservation Society captured four in the wild in 1962 and, including one more from London Zoo and four from King Saud of Saudi Arabia, established a world herd in Phoenix Zoo, Arizona.	In the 1970s the Sultan of Oman pledged to restore the animal in the wild in his country and in the 1980's two herds were re-introduced into the wild. There are now around 100 individuals surviving in the wild, classified as endangered. (Tudge, 1991)

Table 3.6: **Case histories of three different species**

3.3.10 Link Between Exploitation, Development, and Conservation

The rapid rate of deforestation in tropical forests is of particular concern. Many areas of tropical forests are found in Less Economically Developed Countries (LEDCs) and have been an important source of revenue to boost economic development from logging and for commercial farmland for soya bean, oil palm, and rubber plantations. In comparison with deforestation of the temperate deciduous forests of North West Europe, there has been a comparatively sudden impact on the tropical forests which has brought them to public attention. England's native forests in comparison have been reduced to 1% of the countries area but this has taken hundreds of years, not a few decades.

Public awareness of the predicament of rainforests rose greatly during the 1980s in many More Economically Developed Countries (MEDCs). This was accompanied by increasing awareness of global environmental issues and was paralleled by the development of green politics. Green parties emerging at the time of this rising awareness championed the cause of 'save the rainforest' in their manifestos. In 2008 a UN joint initiative focused on Reducing Emissions from Deforestation and Forest Degradation as part of action on climate change. The forests of 64 developing countries have been protected under the REDD agreement, with investment from MEDCs.

3.3.11 Case Study of a Natural Area of Biological Significance: The Seasonal Evergreen Forests of Northern Thailand

Ecology and distribution

The seasonal evergreen forest is found in areas over 1,000 m in elevation which receive over 2000 mm of rain per year. In Thailand this forest type is known from only around 3,500 km^2, or around 5% of the country's land area. The trees included a variety of hardwood species, such as oaks, false chestnuts, and dipterocarp trees, with conifers found in a few high-altitude locations. The forest is cooler and wetter than the lower lying forest and has a greater structural diversity (complexity) than the drier forests of the lowlands (Maxwell, 2004).

Biodiversity and endemism

The seasonal evergreen forests are located in one of eighteen global biodiversity hotspots. These are centres of high biodiversity and endemism. Diversity and complexity of these forests is comparable to the tropical rainforests in Borneo.

	Species	Endemism (%)
Plants	13,500	52
Mammals	433	17
Birds	1,266	5
Reptiles	522	39
Amphibians	286	54
Fish	1,262	44

Table 3.7: **Diversity and endemism**
Source: Adapted from Conservation International. (n.d.). Table 4.8: Diversity and endemism in the hotspot

Threats to biodiversity

Commercial deforestation cleared much of the forest until a nation-wide logging ban in 1989. The seasonal evergreen forests were largely saved by a lack of easy access. Despite the ban there is still deforestation occurring from subsistence agriculture and illegal commercial logging. Small-scale clearance for subsistence farming by hill tribes is still practised in some areas along with commercial cropping of tea, coffee and cabbage. Populations in the rural areas of the region are growing much faster than in the urban areas and the area continues to receive refugees from Burma.

Fires are started in the dry season to clear the forest for a variety of reasons to:
- Allow easy access for hunting
- Clear land for planting of subsistence or cash crops
- Encourage growth of wild mushrooms (against the scientific evidence)
- Show power and imply ownership of territory
- Return nutrients to the soil (against the scientific evidence).

Air quality has been affected by the extensive burning, with particulate matter levels often recorded higher than the national standard of 120 µg/m^3 PM$_{10}$, sometimes over 200 µg/m^3 in Chiang Mai itself due to a combination of fires and urban pollution. Air quality has impacted on the lichen communities (Saipunkaew, Wolseley, Chimonides, & Boonpragob, 2007) and, as well as being a significant human health issue, this will also have an effect on other species of animal.

Local extinctions (extirpations) of the larger mammals and birds due to habitat destruction and hunting pressure are widespread. The list of biodiversity lost (local extirpation) includes the great hornbill (*Bucerus bicornis*), several species of arboreal primates, most of the large herbivores such as wild elephant (*Elephas maximus*), forest cattle such as gaur (*Bos gaurus*) and banteng (*Bos javanicus*). Critically endangered species are found in small numbers including the Indochinese tiger (*Panthera tigris corbetti*), pangolin (*Manis javanica*), and the crocodile salamander (*Tylotriton verrucosus*). All are hunted for alleged medicinal value of their body parts and are sold on illegal markets.

These forests have an important hydrological function as watershed forests and help to maintain dry season stream flow. Due to deforestation and increased extraction from the streams increasingly dry season flow is consistently failing. This may have devastating consequences for the remaining larger mammals that also need the water to survive (Graham, 1998).

Climate change may exacerbate all of the current issues. Assuming continuation of current management practice continued warming will see a decline in the evergreen forests as hotter, drier conditions encourage dry deciduous forests.

3.4 Conservation of Biodiversity

3.4.1 The Arguments for Preserving Species and Habitats

Some of the many arguments that can be given for conserving species are presented in Table 3.8.

Ecological	Life-support service value is present in preserving biodiversity in many diverse areas, for example, watershed protection and a stable climate.
	Ecosystem-support functions occur as species are interdependent. Some species are keystone species, which, if removed from the ecosystem, can lead to many other species becoming extinct.
	In relation to the tropical rainforests the biologist E.O. Wilson concluded that 'we are breaking the crucible of evolution'; we are not just removing species—we are also removing the process of generating them (Wilson, 1992).
Economic	The commercial value of species and habitats can be seen directly in terms of natural capital.
	Many habitats have substantial value for ecotourism.
	There is also unknown value in the potential of the species for agriculture, medicine, and biotechnology. Increasingly this value is found in the genetic diversity as well.
	Opportunity costs exist that make this kind of research expensive; it is not just the cost of researching potential that need to be included but the added cost of not harvesting.
Aesthetic	Nature is both the subject matter and the provider of inspiration for countless works of poetry, literature, and art since the earliest cave paintings (see both the quotes below as examples).
	An aesthetic experience of nature is recognised as significant in healthy childhood development, mentally and physically. (Nature deficit disorder has been proposed as a condition that links to many other learning and behavioural disorders).
	In some mysterious way woods have never seemed to me to be static things. In physical terms, I move through them; yet in metaphysical ones, they seem to move through me. Fowles, 1992
Ethical	Some ethical arguments stress the value of biodiversity to us; they link to the economic/ecological consequences and costs of not protecting biodiversity. These arguments promote the idea of good stewardship (looking after the environment) and sustainable development for the good of future generations.
	Other ethical positions may argue for the intrinsic value of the environment (1.1.7) or rights of individuals or species to exist.
	One impulse from a vernal wood *May teach you more of man,* *Of moral evil and of good,* *Than all the sages can.* 'The Tables Turned' by William Wordsworth

Table 3.8: **Summary of arguments for preserving biodiversity**

3.4.2 International Action: The Role of Intergovernmental and Non-governmental Organisations

The United Nations Environment Programme (UNEP) is an example of an intergovernmental organisation (IGO). The organisation aims to provide leadership and encourage partnership for environmental management through inspiring and informing nations and people. The World Wide Fund for Nature (WWF) is a non-governmental organisation (NGO) and one of the world's largest independent conservation bodies. The organisation aims to stop habit destruction and degradation globally, promote conservation of biodiversity, and ensure sustainable resource use and pollution reduction.

3. BIODIVERSITY AND CONSERVATION

	UNEP (IGO)	WWF (NGO)
Use of media	Media used for reporting progress as news events through press releases. Generally, it does not sensationalise events.	Media is used for fund raising and awareness raising combined. A variety of media are used as part of campaigns that may use sensationalist tactics to gain public support.
Speed of response	Promotes political agreement through the UN and takes time to build consensus.	Independent decision making within the organisation—can make decisions quickly.
Diplomatic restraints	Works through legal frameworks and process in each country.	Works comparatively free from diplomatic consideration but through legal methods in any country.
Political influence	As the UN designated authority, it coordinates international political agreement.	Works to raise political awareness and campaigns on environmental issues. Works within the political framework of the UN during negotiations on conventions.

Table 3.9: **Comparison of the roles of UNEP and WWF in international conservation efforts**

3.4.3 International Action on Biodiversity

The World Conservation Strategy, 1980 (Updated in 1991). The World Conservation Strategy was written by a partnership of three major global conservation organisations: the United Nations Environment Programme (UNEP), World Wildlife Fund (WWF), and the World Conservation Union (IUCN). It aims to:

1. Maintain the planet's life support systems
2. Preserve genetic diversity
3. Support sustainable use of species and ecosystems.

The Global Biodiversity Strategy (1992) by the World Resources Institute (WRI), IUCN, and UNEP which has the same goals as the Convention on Biological Diversity (agreed at the Rio Earth summit, 1992). It aims to:

1. Conserve biodiversity
2. Promote sustainable use of resources
3. Promote fair sharing of benefits from biodiversity.

Distinguish between strategies, conventions and protocols

Strategies: suggested courses of action that have no legally binding agreement.

International conventions: agreements that suggest targets and encourage government action.

International protocols: set targets in a legally binding commitment to which governments agree.

Link these events to the timeline of environmentalism (1.1.6).

The convention encouraged the development of national biodiversity action plans. Further targets were agreed on in the Johannesburg summit (2002), a target was set of 'a significant reduction in loss of biodiversity at the global, regional and local level' by 2010, the International Year of Biodiversity.

In 2010 the Nagoya Protocol set new objectives of promoting the fair and equitable sharing of genetic resources, as an addition to the convention. The role of economic incentives and the review of environmentally damaging subsidies were identified as the key to saving biodiversity.

Going backwards

At Nagoya in 2010 it was accepted that the conventions targets had not been met and the loss of biodiversity continues.

Key objectives and a strategic plan for Biodiversity were developed for the UN Decade on Biodiversity 2011–2020. The key focus of this decade maintained the focus on the role biodiversity has in sustainable development.

A series of strategic goals were established:

1. Address underlying cause of biodiversity loss;
2. Reduce pressure on biodiversity;
3. Improve the status of biodiversity;
4. Enhance biodiversity benefits for all;
5. Enhance implementation.

3.4.4 Habitat and Species-based Approaches

Species-based approaches to conservation mean focusing on a single species with the intent to prevent its extinction. Habitat-based approaches focus on protecting areas of the environment and all species within them. Often conservation charities use a mixed approach for different reasons.

3.4.5 Species-based Approaches

The Convention on International Trade in Endangered Species (CITES) was agreed in 1963 amongst IUCN member countries. CITES runs into problems with enforcement and monitoring. Customs officers are trained to recognise trade in endangered species and charities, such as the EIA (Environmental Investigation Agency), do detective work to follow the trade. However, endangered species are still traded for fur, horns, skulls, and bush meat in many LEDCs and across borders that are difficult to monitor.

Captive breeding and reintroductions can be carried out in an attempt to save species. The example of the Arabian oryx shows how this can be successful and other success stories include the golden lion tamarin in Brazil. In the case of Schomburgk's deer it could have been successful had deer been captured, as deer are easily bred in captivity. Attempts to breed larger animals may be difficult and may result in reduced genetic diversity, as with the South China tiger. Sperm banks and artificial insemination techniques can be used to share global genetic diversity between zoos without transferring animals. Even when suitable habitat is available, animals are not always easy to reintroduce into the wild as a sustainable population. Botanic gardens and seed banks help preserve plant biodiversity in the same way. Seeds harvested from the endangered strapwort at Slapton were collected and sent to Kew Gardens.

Some species have higher aesthetic value, others a more significant ecological role. There are a variety of terms to describe such ecological and conservation roles:

- Charismatic megafauna: large attractive animals that can be used to promote conservation
- Flagship species: by raising one species profile the broader goals of conservation gain more support
- Umbrella species: protect many other species if conserved
- Keystone species: a species that has a significant niche role in the ecosystem; its removal impacts on many other species

Public interest can be motivated easily for species like the giant panda chosen as the WWF logo. Currently, the polar bear is filling the same role for threats to the Arctic ecosystem. Species that may be less charismatic may be more significant in terms of their ecological

3. BIODIVERSITY AND CONSERVATION

value—bacteria and fungi may provide more important functional roles in the ecosystem but who would donate to the 'save the boring brown fungi' fund?

3.4.6 Habitat-based Approaches

Habitats are protected through a wide variety of approaches and legislation. International designations for protected areas include wilderness areas, non-hunting areas, national parks, and nature reserves. Systems vary between countries; UK National Parks include people living, whereas in the US model they are pure wilderness areas. In some wilderness areas indigenous people have been relocated or their activities restricted.

Habitat-based approaches may lead to disagreements on appropriate management regarding people who live or work in the area, or the management of the wildlife. UK national parks, for example, use their influence on farming practice or buildings within the national parks. Some conservationists argue that these managed ecosystems should be allowed to return to a wilderness state, a process called 'rewilding'. Others argue that protected areas need management to protect species and landscape value, along with the livelihoods of people who live there.

3.4.7 Nature Reserve Design

The principles of island biogeography are often applied to nature reserve design. These studies are applied as nature reserves can be seen as islands in a sea of farmland. Islands also exist in natural habitat such as the 'islands' of seasonal evergreen forests in a sea of dry seasonal forests, now further isolated by farmland, settlements, and roads. In Table 3.10, the diagrams show the best reserve on the left-hand side.

Island Biogeography Studies	Application	
Species richness on islands shows an increase related to area. Rates of extinction depend on the size of habitats. Small islands will have a larger extinction rate than larger ones.	Larger reserves are better than small ones.	
Rates of colonisation from the mainland depend on the degree of isolation. The nearer the mainland the faster the habitat will be colonised.	Reserves nearer to other reserves, or connected by wildlife corridors, are better than those further away (isolation is less).	
Wildlife corridors between protected areas allow populations to move.	This may involve protection against hunting or removal of barriers such as fences or tunnels under roads.	
Edge effect* is the term used to describe the changes in conditions found where the reserve meets surrounding land use.	For example, in a forest this area may be hotter and drier than the centre and can support lower diversity. In this case a greater distance from the edge is needed.	

Table 3.10: **Island biogeography applied to reserve design**

In some instances, habitats are managed to promote edge effects between habitats in a reserve; this may be in relation to promoting the amount of 'edge', for example, between reeds and open water.

3.4.8 Case Study: The Success of Slapton Ley National Nature Reserve, UK

Slapton Ley is a 192 ha site located on the South Devon coastline. It contains a variety of habitats, including the largest freshwater lake in South West England, traditional wet meadow, broad deciduous woodlands with coppice, reed marsh, wet woodland, and shingle ridge. Slapton Ley is a Site of Special Scientific Interest (SSSI), which number thousands throughout the UK. SSSIs are classified by Natural England because they contain a high diversity of species or habitats or a rare species or habitat. The reserve is also one of 88 National Nature Reserves (NNR) designated by Natural England. The site is managed by a reserve manager who carries out management according to an agreed plan discussed with all the stakeholders. Slapton Ley is owned by the Whitley Wildlife Conservation Trust and managed by the Field Studies Council. As well as a site known for its biodiversity, it is one of the best researched nature reserves in the country.

The management plan includes a variety of techniques for managing the reserve to protect the important species. Slapton is noted as the only British locality for the plant strapwort, which is a pioneer plant that grows on gravel shores on the lake. Cetti's warblers have a nationally important breeding population in the reed beds. There are over 2,000 species of fungi recorded, including 29 new to science.

Reed fringe	Cattle used to trample reed fringe vegetation for strapwort.
Coppice	Cut on a rotational cycle to promote different ages of woodland stand.
Shingle ridge	Cutting patches on yearly rotations to maintain diversity of ridge plants.

Table 3.11: **Some of the management activities that have been carried out on the NNR**

See Chapter 5 for connection to eutrophication.

The reed fringe management promoted strapwort recovery, but the cattle increased soil erosion and introduced faeces directly into the lake, adding to the nitrate and phosphate concentration. Additional issues for the lake include sedimentation and succession, which will eventually convert the freshwater body into wet woodland. Control of water quality in terms of nutrient input and sediment load involves setting up agreements with farmers in the catchment area. The water quality has been monitored through 45 years of research at Slapton, though this has only resulted in limited influence on the agricultural management of the catchment area.

Clearance of the shingle ridge was a partial success until coastal erosion of the ridge caused inundation and a deposit of shingle on much of the site in the winter of 2001. This has highlighted a major issue for the site; if the shingle ridge breaches, the lake will change to a saltwater marsh.

Otter and mink at Slapton

The otter, *Lutra lutra*, is a native mammal that lives at Slapton Ley and also alongside riverbanks nearby. Sensitive to human disturbance, its populations have been in decline throughout the 20th century. The North American mink (*Mustela vision*) is not native to the area; it was released from fur farms across England by animal rights activists and escapees from abandoned fur farms as the industry lost popularity. Scientists were concerned that there would be intense interspecific competition for food and that the otter may be pushed out by the aggressive invasive species.

Studies carried out on the faeces of the animals establish clearly what their diet is, and it is not necessary to make observations of the animals to find out what they are eating. Otters are specialist fishers while the mink have a mixed diet of aquatic and terrestrial prey. Studies made in different locations have compared the diet of the mink in the presence of otters and without. Studies have also been carried out in the same locations, with changing densities of otters over time. Here is some typical data from a freshwater lake and a river in Devon.

		Percentage in Diet (%)			
		Fish	Amphibia	Birds	Mammals
Slapton Ley	Otter	92	<1	4	3
	Mink	22	<1	37	29
River Dart	Otter	82	5	2	6
	Mink	24	10	4	59

Table 3.12: **Some sample data from habitats where mink and otter coexist**
Source: Adapted from Riley, 1996

Otters are protected at Slapton by designation of core areas with restricted human access. Mink are culled in order to reduce their predation of birds living in the reed fringes.

Practice questions: Biodiversity and Conservation

1. *Outline* the mechanism of natural selection.

   ....
   ....
   ....
   ....

2. *Describe* how different forms of isolation lead to speciation by natural selection.

   ....
   ....
   ....
   ....

3. *Describe* and *explain* the influence of plate activity on evolution.

   ....
   ....
   ....
   ....

4. *Describe* and *explain* the relationship between plate movements and biodiversity.

   ....

   ....

   ....

5. *Describe* and *explain* how:
   (a) Community diversity changes through succession

   ....

   ....

   (b) Greater habitat diversity leads to greater species and genetic diversity

   ....

   ....

   (c) A complex ecosystem may provide stability and its capacity to survive change may depend on diversity, resilience and inertia

   ....

   ....

   ....

   ....

   (d) Human activities alter diversity through modifying succession and simplifying agricultural ecosystems.

   ....

   ....

6. *List* the naturally occurring (non-human) factors that lead to a loss of diversity.

   ....

   ....

   ....

7. *Discuss* the accuracy of estimating extinction rates in the past and present.

   ....

   ....

   ....

3. BIODIVERSITY AND CONSERVATION

8. ***Compare*** background extinction rates with those found in the present day.

   ....
   ....
   ....

9. ***List*** the reasons for extinctions that are caused by people.

   ....
   ....
   ....

10. From the list you have written in the previous question, ***explain*** which you think is the most significant.

    ....
    ....
    ....
    ....

11. ***Discuss*** the perceived vulnerability of tropical biomes and their relative value to global biodiversity.

    ....
    ....
    ....
    ....

12. ***Outline*** the factors used to determine status on the IUCN's Red List.

    ....
    ....
    ....
    ....

13. ***Describe*** the case history of three species:
    (a) Extinct

    ....
    ....
    ....

(b) Critically endangered

....

....

....

(c) Improved by intervention.

....

....

....

14. **Describe** the case history of a natural area of biological significance and threats to it from human activities.

....

....

....

....

15. **State** the arguments used for preserving species and habitats against the headings in the following table:

Ecological	
Economic	
Aesthetic	
Ethical	

16. **Compare** the role and activities of IGOs, such as UNEP, and NGOs, such as WWF in protecting biodiversity in terms of:

(a) Use of media

....

....

....

(b) Speed of response

....

....

....

3. BIODIVERSITY AND CONSERVATION

(c) Diplomatic restraints

....
....
....

(d) Political influence.

....
....
....

17. Targets set at Rio: Promises from 182 countries to abide by the convention and introduce national strategies for conservation. What happened: the two richest biomes, tropical forests and coral reefs, suffered increased damage. Estimates of the human accelerated global extinction rates are showing no signs of decreasing. *Evaluate* the success of international agreements in protecting biodiversity.

....
....
....
....

18. *State* and *explain* the criteria used to design protected areas.

....
....
....
....

19. Look at the data on mink and otter diets. *Suggest* how the data would look different for mink if otter were not present.

....
....
....
....

20. *State* and *justify* your opinion on the culling of mink in the Slapton Ley NNR.

....
....
....
....

21. *Evaluate* the success of a named protected area using an example, such as Slapton Ley NNR.

....

....

....

....

22. **Discuss** and *evaluate* the strengths and weaknesses of the species-based approach to conservation.

....

....

....

....

....

....

Chapter 4: Water and Aquatic Food Production Systems and Societies

4.1 Introduction to Water Systems

4.1.1 Earth's Water Budget

On land there is great variation of water supply due to climatic factors. The driest places are found inland in continental type climates, in rain shadows and in the desert belts where dry air is descending. Coastal regions that are found nearer to warmer ocean currents often have greater precipitation as winds blow in from the sea. Water on the planet is mostly stored in the oceans, with only a small proportion as freshwater. Most of the freshwater is in glacier and ice caps, as shown in Table 4.1.

Saltwater	Oceans and seas		97.2
Freshwater	Glaciers and ice caps		2.15
	Groundwater		0.62
	Freshwater lakes		0.009
	Inland seas	2.8	0.008
	Soil moisture		0.005
	Atmosphere		0.001
	Rivers and streams		0.0001

Table 4.1: **Percentage proportions of the world's water in different stores**
Source: includes data from Wright & Nebel, 2002; and Aquastat 2016

> You don't need to memorise these figures exactly but you should be able to describe the general pattern of distribution of water shown below.

4.1.2 Hydrological Cycle

Water moves though a cycle driven by solar energy, known as the hydrological cycle. It moves water, dissolved matter, and energy through oceans, overland, and through the atmosphere. This cycle can be viewed as a system with water moving through flows and stores as seen in Figure 4.1.

ENVIRONMENTAL SYSTEMS AND SOCIETIES SL

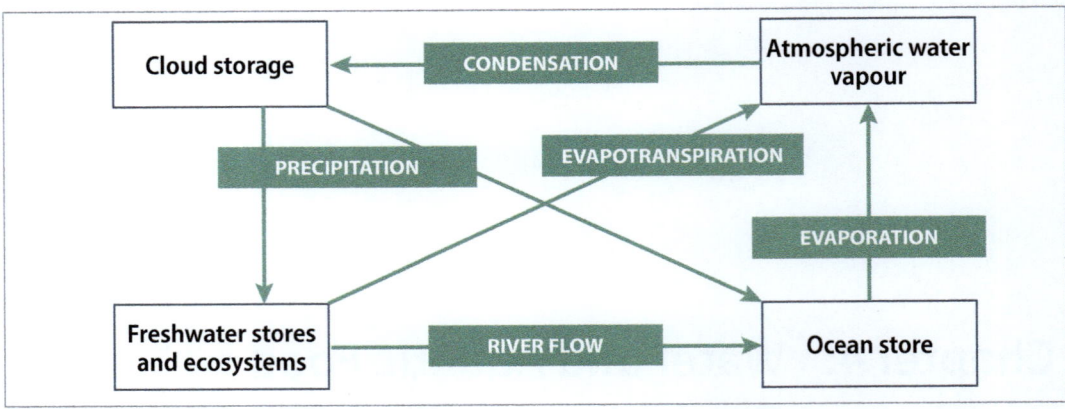

Figure 4.1: The hydrological cycle

You should learn definitions for the key terms in the syllabus shown in tables 4.2 and 4.3

Term	Detail
Atmospheric store	Water vapour in atmosphere supplied by ET from surface.
Condensation into cloud store	Water vapour condenses to liquid water in clouds due to temperature change and the presence of dust or other particles (nucleating surface)
Precipitation to surface	Precipitation falls due to the build-up of water droplets, their combining to greater mass, and then falling as rain or snow.
Water storage	Stores include the bodies of all living organisms, or non-living storage such as atmosphere, ice cap, ground water, lakes, and ocean.
Rivers and glaciers	Flows of water overland as ice or liquid.
Evapotranspiration (ET)	ET is the combination of evaporation and transpiration; it requires both heat from sunlight and the biological action of plants
Convection	Water carried by hot, moist rising air higher up in the atmosphere.
Advection	Wind-driven movement of water horizontally through the atmosphere
Sublimation	Snow and ice moving directly to atmospheric storage from solid to gas

Table 4.2: Terms used to describe the hydrological cycle

4.1.3 Overland Movement of Water

The movement of water between stores connects soils, agriculture, and water resources, as shown in Figure 4.2. Human management of the land and soil influences the water quality, including agricultural practices, deforestation, and urbanisation.

4. WATER AND AQUATIC FOOD PRODUCTION SYSTEMS AND SOCIETIES

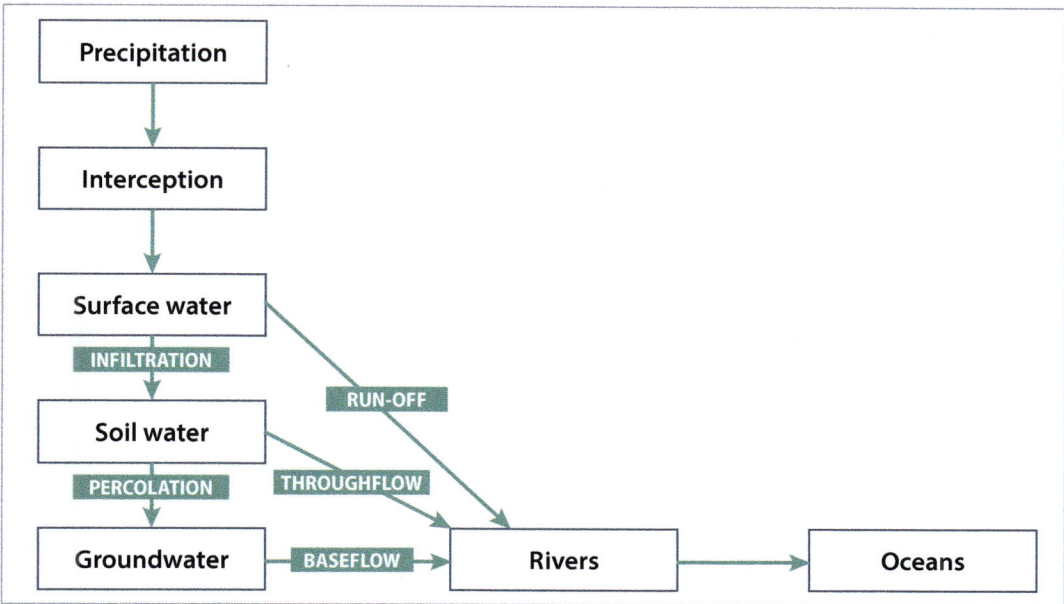

Figure 4.2: **The overland flow of water**

Term	Detail
Interception	Precipitation may be intercepted by vegetation or hit the ground.
Surface water	Water storage on the ground surface.
Run-off	Movement of water over the ground surface to rivers.
Infiltration	Movement of water from surface through the soil.
Percolation	Movement of water through porous rock and sediment.
Throughflow	Movement of water through soils to rivers.
Groundwater	Storage of water in rocks underground (aquifers).

Table 4.3: **Terms used to describe the overland flow of water**

Water carries dissolved materials, including organic and inorganic pollutants, through the cycle. Surface run-off and water bodies may carry insoluble materials as well. Reducing infiltration rate through human activity can cause increases in surface run-off, impacting on water quality and increasing flooding.

4.1.4 Oceanic Circulation Patterns

Ocean currents help to distribute heat around the world, essentially moving hotter water from the tropical regions to higher latitudes. Warm ocean currents move water considerable distances such as the Gulf Stream that flows from the Gulf of Mexico to Scotland in the UK. Cold water currents move from high latitudes to lower latitudes, such as the Peru Current that carries cold water from the Antarctic Ocean along the west coast of South America.

Density differences also arise due to differences in temperature and salinity.

- Cold water is denser than warm, and so cooler currents run deeper in the ocean.
- Freshwater is less dense than salty, so water with lower salinity stays close to the surface.

The circulation pattern produced due to these factors is known as thermohaline circulation.

Thermohaline circulation is a driving force of the ocean conveyor belt that connects the major ocean basins of the world. The ocean conveyor belt is important in determining both marine and land biome types due to its influence on nutrients in the water and climate. The cold water currents run between two to four kilometers beneath the water surface.

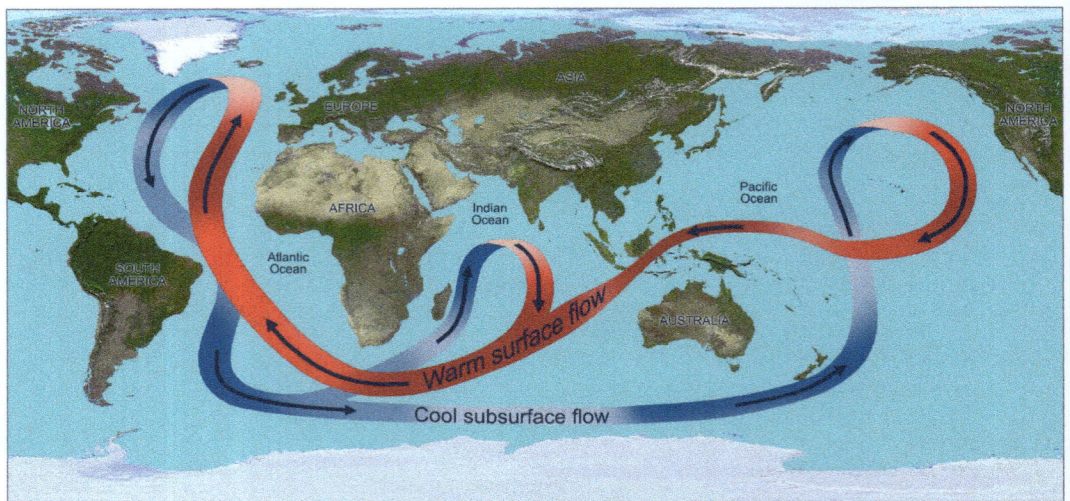

Figure 4.3: **Figure 4.3: Schematic diagram of the ocean conveyor belt**
Source: NASA/JPL

4.2 Access to Freshwater

4.2.1 Availability of Freshwater

Water availability is naturally varied due to variations in climate and geological formations. Patterns of precipitation are discussed in Chapter 2, under biomes. Patterns of drainage can produce large variations in water; some rocks allow water to permeate through and are called permeable. Others are impermeable to water, preventing water going deeper into the ground. Porous rocks are useful water stores, and provide space for large amounts of groundwater. These may form important aquifers that supply water to many urban and agricultural regions. This water may have formed from precipitation a long way from the site, or from the geological past (fossil water). Surface water bodies and channels may store large amounts of freshwater.

4.2.2 Access to Freshwater

Access to freshwater varies greatly for economic reasons and due to infrastructure development. MEDCs often have piped water that is drinking quality in houses, even to flush toilets. In LEDCs this is less common and piped water may not be available, or when it is it may not be potable. Drinking water may be purchased in large bottles, delivered to the house by those that can afford it. Access to good sewerage also varies considerably with development and economy, which can impact on the quality of drinking water.

Economic limitations that influence access to freshwater occur where piped or bottle water is paid for. The costs of piping water for recreational use, for example, in swimming pools or extensive gardens, may be exorbitant for many individuals in a society. In some parts of the world social conflict may arise where finite water supplies are extracted for use by exclusive sectors of the society; for example, golf courses may be in competition with domestic usage in villages or agriculture. Tourism can cause significant impact in some areas, where water availability is naturally low, for example, tropical islands or semi-arid areas. Tourists use disproportionate amounts of water in many LEDCs, over 15 times greater usage in locations such as Goa, Zanzibar, and Bali. Tourists in Bali may use over 3,000 litres per day, whereas local consumption is less than 100 litres per day (Noble, Smith, & Pattullo, 2012).

4.2.3 Climate Change

Climate change has been predicted to have a significant influence on water availability in the future. There are already observable effects of changes in water distribution that can be linked to climatic change with current warming. This is due, in part, to projected changes in precipitation patterns and rates of evapotranspiration. These changes further indicate that arid areas are moving to higher latitudes as the tropical cells increase in size. Savanna type conditions are also predicted for areas that are currently rainforest, such as the Amazon basin.

4.2.4 Increasing Demand

Water demands are increasing rapidly on a global scale; freshwater use has tripled in the last 50 years. This increase can be partly explained by an increase in population growth, but water consumption is growing faster than population. In part, this can be explained by development stage: as countries develop their populations they use more freshwater. The difference in these trends is summarised in Table 4.4, with some possible reasons given.

% Water consumed	World	LEDCs	MEDCs	Some reasons for the difference
Domestic/urban (Washing, cooking, drinking, waste disposal)	12	4–15	15–25	Increased personal usage from lifestyle change, for example, dishwashers, baths, swimming pools, piped water, and sewerage.
Industrial (Waste disposal, solvent, coolant, and a raw material)	19	5–15	35–75	Development of water hungry industries, for example, steel production, fertiliser manufacture, and cooling for electricity production.
Agricultural (Irrigation, livestock drinking, cleaning and waste disposal)	69	70–94	5–35	LEDCs often have more basic irrigation technology, which may be prone to water loss by evaporation or leakage. MEDCs are more likely to have supply on demand, and better control over the water supply.

Table 4.4: **Typical values of water use by sector**
Source: Adapted from AQUASTAT, 2016

There are many exceptions to these general trends, due to the patchy (unequal) distribution of water availability, for example, Singapore and Bahrain have very little freshwater. How water is used can also vary considerably, for example, the USA has a large proportion of water used for agriculture.

ENVIRONMENTAL SYSTEMS AND SOCIETIES SL

4.2.5 Contamination and Unsustainable Usage

Water resources may become contaminated by pollution—examples are described in section 4.4.

Water usage may become limited through unsustainable extraction. Simply put, if the outputs (water extraction) is greater than inputs to the system (precipitation or other), then the stock will decline.

Location	Inputs	Outputs	Summary
Ogallala aquifer, USA (detailed case study in 5.3.5)	Aquifer recharges slowly through rainfall, snow melt in the Rocky Mountains and seasonal wetland ecosystems.	This aquifer is one of the world's largest and provides water for over 25% of irrigated land in the United States.	The aquifer continues to decline at rates of over 80cm a year on average and estimates give the aquifer a future life span of only 25 years.
Aral Sea, Kazakhstan, and Uzbekistan	River and ground water input declined by over 60% in 100 years. Water is used to irrigate surrounding land for cotton and other agriculture.	Mostly evaporation, as a saline sea it is not used directly for irrigation. Total evaporation now low, as most of the area is desert.	The formerly fourth largest inland sea, which has now declined by 95% of its area.
Tuvalu, Polynesia	Well water is no longer used for drinking due to poor water quality. Rainwater harvesting is the main source of water, which also reduces recharge of the aquifer.	Digging deeper wells and pits has broken through the freshwater layer in places. Salt water incursion followed over extraction in 1999 and 2000.	Groundwater is regularly contaminated by saltwater at high tide and ecological effects have been observed.

Table 4.5: **Three examples of unsustainable water abstraction**

4.2.6 Sustainable Water Supply Strategies

There are a wider number of more sustainable strategies that can be used to manage water supplies, including:

- Reservoirs
- Redistribution
- Recharge of aquifers
- Rainwater harvest
- Desalination

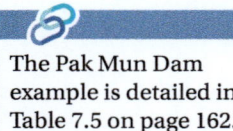

The Pak Mun Dam example is detailed in Table 7.5 on page 162.

Water can be stored in reservoirs, which can help control the water supply with benefits in preventing downstream flooding and conserving water for drier times. Issues arise due to conflict locally over managing the water, for example, the Pak Mun Dam in Thailand. In this case, the fisherman wanted water to flow downstream to allow fish to breed, the farmers wanted to extract it for irrigation, and the Electricity Generating Authority of Thailand (EGAT) wanted to use keep the water supply behind the dam to drive their turbines.

Other systems may be used including redistribution of water from a region with high water to one with low water availability, for example, China redirecting water from the south to the north transport water from the Yangtze to the Yellow River. Redistribution schemes can also be used to recharge aquifers. Rainwater harvesting schemes allow water to be collected in times of plenty, for example, from rooftops, and stored for when it is needed. This is best shown by the white ridged roof and underhouse storage designs required by law in the water stressed islands of Bermuda.

Desalination plants are particularly of value in areas that have little access to freshwater, but plenty access to saltwater and cheap energy supplies. In some regions there has been rapid growth of this approach since the 1980s, for example, United Arab Emirates and Saudi Arabia. Techniques such as cloud seeding (rain making) may be used to improve ground water supplies. Careful management of catchment areas (watersheds or drainage basins) can help keep a more regular supply of water in rivers, even during drier seasons.

4.2.7 Domestic Water Conservation

Domestic water conservation can help reduce excessive demand. For example, most toilet flushing is used to remove liquid waste and needs much less than 8 litres to remove it. Water conservation toilets, designed to flush with a light or heavy flush, save a lot of water, as does putting a few bricks into the toilet cistern of older designs. Traditional squat design toilets with a hand flushing bucket use less water than flushing toilet; some traditional designs of squat toilet separated solid from liquid waste at source.

Figure 4.4: **Composting toilet on an environmental protest camp in London (1996)**

Plumbing systems can be designed to direct grey water (wastes from sinks and showers) into the toilet system or for use in the garden. Water contaminated by faeces or urine in the right drainage conditions can exit via a land drain where the nutrient can be recaptured for plant growth which can filter the water, for example, reed bed filtration systems. Sewage can be treated effectively before it enters the water to kill pathogens or even prevented from entering the water by using dry compost toilets. Figure 4.4 shows a dry composting toilet built as part of a permaculture camp highlighting the misuse of urban land in London (Wandsworth Eco-Village 1996, The Land is Ours). Alternative toilet systems often require a change in attitude to be acceptable to many people.

What do you think? How many Londoners would be happy to use the toilet design shown in Figure 4.4?

4.2.8 Case Study of Water Resource Conflict: Turkey, Syria, and Iraq

Turkey is rich in water resources, with 25 river basins and over 140 HEP plants. Water from Turkey flows across the border, particularly through the Tigris and Euphrates rivers, into neighbouring Iraq and Syria. Syria receives water in the Euphrates from Turkey; Iraq is downstream of both countries. There has been international conflict over these resources between all countries since 1975. Iraq suffers the worst consequences, and the decline of flow in the Tigris and Euphrates has led to salt water incursion from the Persian Gulf along the Shatt Al-Arab river. Syria completed a major dam in 1975, also when Turkey began dam construction as part of a large hydroelectric project (South East Anatolian Project), including the projected construction of around 20 HEP plants and associated dams. Iraq and Syria have both been drying out, causing problems with both food production and social stability as farmers abandon fields. There have been a number of conflicts over water resources between the three countries. Climate change has predicted a reduction in precipitation in this region, which has been experiencing a number of major droughts lasting for many years at a time. Here are some key dates.

1975	Syria reduced water flow into Iraq on the Euphrates when it completed the Tabqa dam, almost leading to armed conflict. Syria closed its airspace to Iraq and troops were mobilised in both countries.
1987	Turkey and Syria came to an agreement that in return for Turkey ensuring flow rates in the Euphrates then Syria would help deal with the Kurdish rebels living in Syria.
1990	Completion of the Ataturk Dam in Turkey resulted in a major flow reduction of the Euphrates, with temporary flow reduction as high as 75% as the dam filled. Iraq threatened to blow up the dam. Turkey retaliated by saying that it would cut of the water flow completely to Syria and Iraq. The dam is declared a weapon of war by Iraq and Syria.
2006–10	Extended drought in region and crop failure in Iraq.
2014	Drought, conflict and the rise of Islamic State, who also take control of the Tabqa dam
2017	US and Syrian forces launch attacks to regain control of the Tabqa Dam.
2017+	Ilisu Dam completion on the Tigris River; this project has been nearing completion for some time and has been postponed on a number of occasions.

4.3 Aquatic Food Production Systems

4.3.1 Some Aquatic Food Production Systems

Aquatic food production systems are traditional in many parts of the world, but growing population and changing technology has also led to newer methods. In aquatic food production systems, the base of the food chain is often phytoplankton, floating primary producers that support a range of food webs. Seaweeds and coral also grow in shallower waters; they are similar to plants and carry out photosynthesis, as do the symbiotic algae that live inside the corals. Both algae and animals are harvested as food resources by humans. Often, carnivorous fish from the top of the food chain are harvested, such as tuna or salmon. Other important species such as cod are further down the food chain. Mariculture systems that are used to cultivate oysters or algae (seaweeds) for consumption are more efficient systems than open ocean fishing for carnivorous fish.

4. WATER AND AQUATIC FOOD PRODUCTION SYSTEMS AND SOCIETIES

4.3.2 Variations in Productivity

Considerable variations are found in the quality of ocean fisheries. The best fisheries are the shallower continental shelves, upwellings, and coral reef areas. Around coastal areas seaweeds and corals can grow up to a depth of 30–40m allowing higher productivity. Open oceans have low productivity. The reasons for these differences relate to the variations in biomes described previously in Table 2.8.

4.3.3 Ethical Considerations

There have been conflicts of interest regarding rights to harvest marine resources and rights of the animals. In particular, the hunting of mammal species such as whales and of seals have seen conflicts over ethics and rights. One example of this is found in the Pacific Northwest of the USA, where the Makah tribe based its cultural existence greatly on the Gray Whale. Their hunting practice links to many traditions and rituals. Hunting of these whales stopped for most of the 20th Century, and was banned under international agreement in 1946. The Gray Whale is no longer endangered and so the Makah have re-asserted their whaling rights, though only one whale a year. They resumed hunting in 1999 a move that is strongly opposed by marine conservation groups such as Sea Shepherd.

4.3.4 Unsustainable Fishing

Fish populations may be overfished by simply catching too many fish, and not allowing the fish to naturally regenerate their populations. There are many examples of overfishing in the world such as shark populations in the Andaman Sea, cod in the North Sea or the Newfoundland Grand Banks, or Atlantic bluefin tuna (*Thunnus thynnus*). Overfishing can lead to the collapse of fisheries leading to total loss of viable stock. In the English Channel pilchard fisheries increased yield throughout the 19th century, but now there are no longer pilchards found in many of the old fishery areas in Devon. Unsustainable fishing normally refers to fishing that occurs at a level that is above the natural reproductive rate for the population. The unnatural selection pressure exerted on the evolution of the fish can also be considerable, leading, for example, to the dramatic decrease in the size of breeding cod in the North Atlantic. Here female cod have evolved to breed at a size small enough to get through fisherman's nets.

Fishing too much is just one example of a number of ways in which the marine environment is damaged. On coral reef some, normally illegal, techniques are practised that damage the habitat and much of the marine community that is not being harvested. Examples include dynamite fishing, poisoning reef fish with cyanide, and the use of very fine nets. Tourist uses of the reef can also cause damage such as the technique of 'sea walking'. In open water huge nets are set up that are sometime referred to as 'walls of death' as they kill all that enters into the net. There is rarely good regulation for these international waters. The technique of drag netting in shallow seas uses weighted nets that not only catch fish but much of what lives on the bottom, and normally causes serious damage to the entire ecosystem.

4.3.5 Policies and Legislation on Fisheries

The 1982 United Nations Convention on the Law of the Sea (UNCLOS) states that 'Territorial Waters' are defined as extending 22 km from the shore of a coastal country. These waters are owned by the coastal countries. The Exclusive Economic Zone (EEZ) is 370 km from the shore. Nations have the rights over the resources and economic activity and are responsible for the environmental protection of this area. This leaves 40% of the world's seas in the control of coastal countries and 60% as international 'High Seas'.

In the European Union there is additional regulation over how much a fishing boat is allowed to harvest within European Waters. The system used is one of quotas that are tradable quantities in the amount of fish that can be caught. As well as regulating catch, the regulation of bycatch is important; this includes all that is unintentionally harvested by fishing practices. There are also other restrictions on the mesh size of nets that can be used, to prevent the capture of immature fish and allow breeding adults to survive. However, policing fishery laws may be difficult and bycatch that breaks regulation may die before being returned to the ocean.

4.3.6 Changing Consumer Behaviour

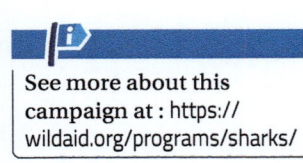
See more about this campaign at : https://wildaid.org/programs/sharks/

Consumers may help as awareness is raised by NGOs to stop consuming certain species, for example, the NGO WildAid's campaign against shark fin soup. Other awareness campaigns have run to protect non-target species; this had led to labelling such as 'Dolphin Friendly' tuna. Often tuna had been caught in nets along with unwanted dolphin which died and were thrown away with other bycatch. WWF have targeted consumers directly, bringing their guide to sustainable seafood to popular restaurants and sharing it with diners to help them choose carefully.

4.3.7 Establishing Maximum Sustainable Yield

Sustainable yield can be calculated from suitable equations, where the growth and recruitment of fish to the adult population from the larval stages represents the input. Death and emigration are outputs from the population.

sustainable yield (SY) = (annual growth and recruitment) − (annual death and emigration)

The difference between the two will be population growth, and, if this is not exceeded, the population should remain steady. However, the maximum sustainable yield of fishery is generally compared to the potential for an area, which may be higher than a given population at a moment in time—particularly if the stock has been over-fished. This is easiest to understand graphically.

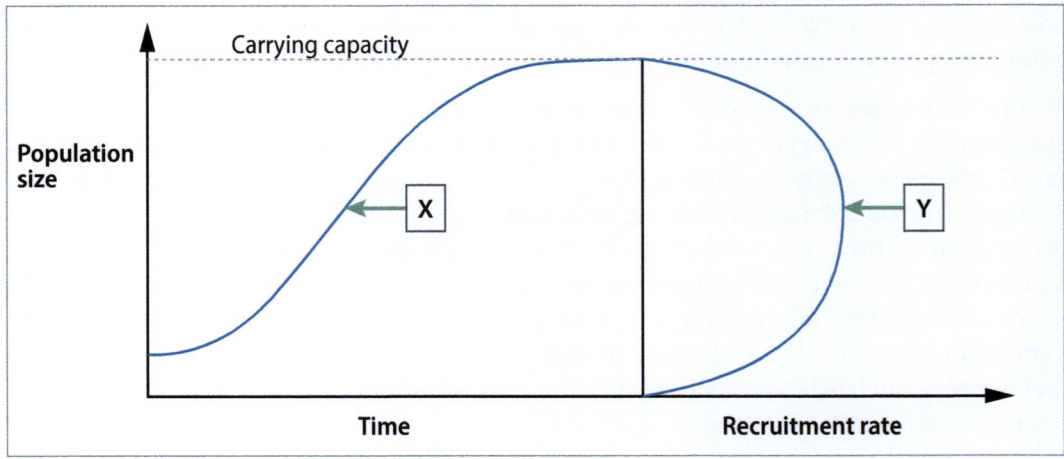

Figure 4.5: **Theory of maximum sustainable yield**

Maximum sustainable yield (MSY) is theoretically at the most rapid growth rate of the population, point X on the diagram. The second part of the diagram shows recruitment to the fish population (growth from immigration and reproduction). On the recruitment curve, point Y represents the maximum sustainable yield. Above this point yield is lost to competition; below this point the reproductive capacity is low and reduces yield.

4. WATER AND AQUATIC FOOD PRODUCTION SYSTEMS AND SOCIETIES

Establishing MSY in reality can be far more complex, as nature may not follow simple projections as complex ecosystem responses and cyclical behaviours may occur.

4.3.8 Case Study Showing the Controversial Harvest of a Named Species: The Grand Banks Cod (*Gadus morhua*) Fishery

This case study shows how controversial actions link to the management of a fishery. The history of the Grand Banks is described in Table 4.6, showing unsustainable fishing levels and collapse.

Decade	Total Catch	Total Fishing Effort and Control
1960s	Rises from 400,000 metric tonnes to 800,000 by the end of the decade.	Number of boats increases from just fewer than 100 to over 300.
1970s	Catch falls dramatically to under 30,000 by the end of the decade.	Number of boats between 400 to 500 vessels. Subsidies help maintain high fishing levels; the establishment of EEZ gives Canada total control over the fishery.
1980s	Catch rises slowly from low in the late 70s to around 350,000 tonnes.	Department of Fisheries and Oceans (DFO) recommends 125,000 tonnes maximum sustainable yield. Boats fall to around 250, then rise to over 500 by end of the decade.
1990s	Yield drops to critically low levels. Cod declared commercially extinct; the Grand Bank closes.	DFO recommends cod fishing stops; total moratorium is imposed, allowing some recovery. Harp seals are blamed and hunted to reduce cod predation.
2000s	Fishery closes	Cod slowly recover, but remain a fraction of original population. Seal kill quotas increased by Canadian Government to reduce predation pressure.
2010s		Studies show cod is continuing to recover, but still around 25% of the 1960s fish population.

Table 4.6: **History of the Grand Banks cod fisheries of the coast of Newfoundland**

Some areas of controversy are outlined as follows:

1. Fish stock surveys and setting maximum sustainable yield is politically controversial.
2. Control over fisheries and unsustainable practice directly impacts livelihoods of 40,000 people.
3. Ethical issues arise over ecosystem management and impact on other species such as culling the harp seal.

Several issues arise, some to do with data reliability and political bias, for example, who is collected and recording the data. The data collection in Newfoundland was seen as controversial by fisherman. In this case, the Department of Fisheries and Oceans (DFO) advised that 125,000 tonnes was the MSY, but the Minister of Fisheries raised it to 235,000 tonnes in 1989 saying it was too low. Environmentalists warned that the DFO level was already too high. Fisherman objected that the survey was too low and that the scientists did not catch the fish effectively, and looked for fish in the wrong places (due to random sampling). Greenpeace have opposed the harp seal hunt, arguing the seals are being scapegoated for poor fisheries management.

4.3.9 The Role of Aquaculture in Supplying Food

The production of farmed fish and shellfish has grown rapidly in recent decades, with aquaculture now supplying over half the fish for human consumption. Land-based aquaculture produces five times the amount produced by inland fisheries, whereas ocean-based aquaculture produces around one third of the wild catch. Without aquaculture, fish for human consumption would have stayed static since the late 1980s, but, instead, due to aquaculture it has continued to rise. It could be assumed that this will simply reduce pressure on ocean fisheries; however, the opposite is true for some types of aquaculture where large inputs of wild fish are used for feed. Aquaculture can also have other ecological effects, including damage to the local habitats, wild collection of seedstock and damage to freshwater quality in the surroundings. Some of these impacts are summarised in Table 4.7.

Type of Fish Farming	Impact
Fish farming	Eutrophication from effluent, invasive species or GMO escapes, genetic contamination of wild stock, spread of disease, and antibiotics can all impact wild fish populations. Use of wild seed stock can impact on further reduce wild populations. Gathering of food such as sand eel produces food web interactions.
Prawn farming	Habitat damage, for example, mangrove forest clearance and coastal wetlands, and loss of ecosystem services. Loss of fish nurseries and impact on wild fish populations.
Mollusc farming	Habitat damage, for example, displacement of seagrass beds, though, this can also enhance biodiversity by increasing habitats for some species. Release of invasive non-native stock spread of disease.

Table 4.7: **Some environmental impacts of aquaculture**

Aquacultural systems vary in terms of their energy efficiency of conversion of food to meat, depending on species. The larger the number, the less efficient the feed to meat conversion, here 1.2 kg of feed is required to produce 1 kg of salmon for example. Remember that the food may be supplied from wild catch fisheries.

System	Animal	Efficiency Ratio (kg of feed/kg of total live weight)
Aquatic	Atlantic farmed salmon	1.2
	Tilapia	1.6–1.8
	Freshwater prawn	1.65
	Saltwater prawns	1.2
	Carp	1.5

Table 4.8: **Feed conversion ratios for aquatic systems by animal type**
Source: Data from Tudge, 2002; Naylor et al., 2000; and National Research Council, 2010

It should be remembered that efficiency is not the only comparison relevant to understanding which systems are more sustainable. For example, prawn farming has led to the destruction of 30% of the world's coastal mangroves and this can lead to a decline in fishery productivity on surrounding coral (see Table 4.6).

4.3.10 Case Study Showing the Impact of Aquaculture: Norwegian Salmon Production

Norwegian fish farming is now a major industry, in particular, of the Atlantic salmon (*Salmo salar*) which earns Norway over 3,000 million US dollars a year. Norway farms around one third of the world's farmed salmon, with other main locations in Canada, USA, Scotland, and Chile (Liu, Olaussen, & Skonhoft, 2010).

Salmon are farmed firstly in a freshwater hatchery, then in a seawater cage. In the freshwater stage, the eggs are hatched and the fish are fry fed until they are large enough to transfer to the sea cages at around one year old. The seawater cages are essentially then used to increase the fish size to market demands of around 3–5 kg.

There are a number of impacts on the environment from this industry, as outlined below:

- Effluent from tanks: this contains organic material from fish faeces, in particular, carbon, nitrogen, and high levels of phosphorous. The phosphorous is particularly important in causing eutrophication problems such as blue-green algal blooms. Blue-green algae are nitrogen fixing, so this has less effect. This is not seen as a problem in Norway as there is plenty of dilution for the effluent.
- Medicines in effluent: this can include antibiotics, disinfectant, and parasiticides. These can have negative effects in the environment including building resistant pathogens and parasites. Other negative impacts include effects on non-target species.
- Escaped salmon: farmed salmon are selectively bred from a small number of wild stock, so their genetic diversity is low. Escapees are unlikely to survive long in the wild, but they do escape and interact with wild stock. They may breed, producing hybrids with low genetic diversity.
- Spread of disease: either from the escaped salmon or from the effluent, diseases and parasites can transfer into the wild stock. In Norway, salmon lice and viruses have been a particular problem for the wild stock. Bacterial disease has been imported from Denmark by fish farmers.

Practice questions: Water and Aquatic Food Production Systems and Societies

1. *Describe* the distribution of water on the planet.

....

....

....

....

2. **Define** the following terms as used in the hydrological cycle:

Term	Definition
Atmospheric store	
Condensation	
Precipitation to surface	
Water Storage	
Rivers and Glaciers	
Evapotranspiration (ET)	
Convection	
Advection	
Sublimation	
Interception	
Surface water	
Run-off	
Infiltration	
Percolation	
Throughflow	
Groundwater	

3. **Draw** a systems diagram of the hydrological cycle and discuss the human impact on the cycle.

4. ***Outline*** some examples of unsustainable water use. ***Suggest*** and ***evaluate*** strategies that can be applied to each of your examples.

....

....

....

5. ***Discuss*** how international disputes over water resources can lead to conflict with reference to a named example.

....

....

....

6. ***Outline*** the history of fishery that has caused controversy. ***List*** some of the controversial issues and ***suggest*** strategies that could be applied to improve the sustainability of the fishery.

....

....

....

7. ***Evaluate*** the claim that aquaculture may be saving wild freshwater and marine species from extinction. Illustrate your answer with details from one case study.

....

....

....

4.4 Water Pollution

4.4.1 Types of Freshwater and Marine Pollution

Human activity generates a diverse range of pollutants that influence water. These can be classified by the sector of society that generates them. The major sectors include domestic, industrial, manufacturing, and agricultural systems. Sources of pollutants are summarised in Table 4.9.

	Type of Pollution	Cause	Effects
Agricultural	Salination	Irrigation	Plants die and this impacts on agriculture and societies food production
	Pesticides	Spraying of crops	Pesticides enter hydrological cycle and food chains, including humans
	Fertilisers, manure, or silage	Spreading fertiliser on fields; manure and silage wash down slopes	Organic or inorganic pollution reduces oxygen in water courses and causes eutrophication
Industrial and Manufacturing	Toxic, e.g., heavy metals, spills, dumping, and leaks	Disposal of by products and waste	Various, can be fatal, e.g. Love Canal
	Thermal pollution	Produced when water is used a coolant for electricity production or industrial processes	Can cause negative impacts on surrounding biodiversity
Domestic	Sewage	Waste from toilets and drains	Can contaminate groundwater
	Solid waste (rubbish/garbage)	Waste in land fill sites	Can contaminate groundwater and coastal waters
Transport	Road run off (chemical cocktail)	Drainage from roads	Can contaminate groundwater and coastal waters
Others	Plastics	Plastics may move as floating waste on the surface of the water, microplastics may move in suspension from domestic drainage	Plastics are varied in chemical form, but many are highly persistent pollutants with a long life. Currently building up in ocean gyres and on remote beaches in
	Light	Surrounding light and noise pollution may impact on aquatic ecosystems	Excessive light can disrupt turtle nesting or produce algal growth
	Radiation	Water used as coolants in nuclear power plants	Can cause tissue damage, mutation and cancer in living organisms

Table 4.9: **Types of water pollution**

4.4.2 Monitoring of Water Pollution

Water pollution can be monitored using multiple parameter tests to indicate generate a water quality index. A 9-parameter test normally includes oxygen concentration, biochemical oxygen demand, temperature, turbidity, phosphate, nitrate, faecal coliform bacteria (human gut bacteria), pH, and total dissolved solids.

4. WATER AND AQUATIC FOOD PRODUCTION SYSTEMS AND SOCIETIES

4.4.3 Oxygen Concentration and Biochemical Oxygen Demand

Oxygen concentration is often used as a measure of organic pollution in water. Direct monitoring of oxygen levels in water can be assessed using either chemical techniques or data logging probes. The chemical test is called the 'Winkler test' and may be carried out using a field kit or by fixing the sample with chemicals in the field and determining oxygen concentration in the laboratory. Data logging probes allow for streamed data to be collected at different locations. It is usual to collect temperature and altitude data, as these influence oxygen solubility and are needed to calculate percentage saturation.

Biochemical oxygen demand (BOD) is a measure of the total demand for dissolved oxygen by the aerobic activity of both living and chemical components in a water body. It is normally measured over a five-day period. Initial oxygen concentration is taken from a water body and samples are collected in sealed bottles with no air spaces. The samples are kept in a dark cupboard for five days. The oxygen content is then measured again and the difference calculated in mg/l.

This method can be used to assess organic pollution levels in water. The demand for oxygen comes from the microscopic community of bacteria and fungi that live in the water. These decomposer populations are much higher where there is sewage released into the water. There is also a chemical demand for oxygen in water. For example, the chemical ammonia demands oxygen to convert to nitrite and nitrate (see 2.3.10 on page 58 for the nitrogen cycle).

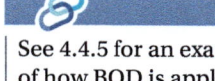
See 4.4.5 for an example of how BOD is applied to assess pollution levels.

4.4.4 Biotic Indexes

Species indicate environmental conditions due to their tolerance limits. Furthermore, if they are present in the environment, then they indicate the conditions at that site for as long as they have lived there, including all seasonal variations. Certain species are termed 'indicator species' as they indicate conditions on a site. Freshwater invertebrates are used to indicate water quality in relation to oxygen concentrations and organic pollution. A variety of methods exist to calculate values for biotic indexes, using the species found in a site. Individual species are given a score and averages can be taken from a species list or related to the abundance of organisms. Diversity measurements can also be used to compare sites.

4.4.5 Comparing Water Quality Upstream and Downstream of a Pollution Source

Locations	Catchment area	Elevation	Dissolved oxygen (mg/l)	Dissolved oxygen after 5 days (mg/l)	BOD in 5 days (mg/l)
Site 1	Forest	1000 m	8.5	3	5.50
Site 2	Forest, agriculture, houses	400 m	7.0	0.5	6.50

Table 4.10: **Comparison of sites sampled BOD measured above and below a pollution source**

The stream at Site 1 drains a forested catchment area. It then flows down through a village and drains an agricultural catchment area, including rice fields and cow farms. Site 2 is below the village and includes drainage from domestic sources of sewage and kitchen waste. There is also a ford through the stream above the sample point. Kick samples were taken from the stream bed. Freshwater invertebrates are kicked into a net with standard size and effort of kicks. Results shown in Table 4.10 are from five kick samples at each site.

Here the invertebrates are scored according to tolerance of oxygen concentrations. For example, stonefly nymphs have a high minimum tolerance for dissolved oxygen. Consequently, they are only found in the first site. Swimming mayfly nymphs are found in both sites but numbers are lower in the first site. The biotic index for each site has been calculated; this is the average score for each site based only on the species presence (for this index abundance is ignored).

Link to Chapter 2, to review niche theory (2.1.6).

Species Recorded	Score	Site 1	Site 2
Stonefly nymph	10	3	
Flattened mayfly nymph	10	1	
Stonefly nymph	10	4	
Spiral stone cased caddis	10	4	
Straight stone cased caddis	10	2	
Leafy cased caddis	7		1
Caseless caddis (orange head)	7	1	
Caseless caddis (brown head)	5	15	1
Saucer bugs	5	1	8
Beetle larva	5	1	1
Other fly larva	5	4	3
Swimming mayfly nymph	4	22	75
Midge larva (bloodworm)	2	1	1
True worms	1	2	4
	Total population size	47	93
	Simpson's diversity	3.01	1.52
	Biotic Index	6.5	4.6

Table 4.11: **Freshwater invertebrates above and below a pollution source**

4.4.6 Eutrophication

Eutrophication is the process of nutrient enrichment of an ecosystem. This term is most often used in reference to bodies of freshwater in relation to inputs of nitrate and phosphate. Eutrophication occurs in lake ecosystems as part of the natural ageing of a lake and its catchment area but it is often accelerated by human activities such as agriculture, road building, and domestic inputs. The process of eutrophication occurs alongside two related process of ageing in the ecosystem: sedimentation and succession.

4. WATER AND AQUATIC FOOD PRODUCTION SYSTEMS AND SOCIETIES

Eutrophication occurs in a series of connected events and involves many feedback loops. The basic sequence is as follows:

- Increase in inputs of nutrients entering the lake from the catchment area
- Increase in plankton (algae) productivity in the lake
- Algal blooms (extensive populations of algae growing on lake surface)
- Increase in total productivity and community biomass
- Increase in dead organic matter from death of above
- Increase in biochemical oxygen demand
- Decline in oxygen levels.

During the process several feedback loops are likely to be set up as shown in Figure 4.6 on page 125.

4.4.7 Impacts of Eutrophication

The impacts of eutrophication may include the death of fish that require high oxygen levels. Also, other organisms may be poisoned due to toxic algal blooms such as the *Microcystis flos-aquae*, a blue-green algae which produces cyanide. Turbidity (cloudiness) of the water increases and this may lead to a loss of large plants on the lake bed that aerates the water.

Although, in the short term, eutrophic conditions increase productivity, eventually they can lead to declines in biodiversity and food chain length. Impacts on local communities include the decline of fisheries and the loss of water resources for people, livestock, and recreation. Bad smells may result from decomposing matter and can impact on tourism and recreation. Dead zones may develop.

4.4.8 Dead Zones

Dead zones are regions of water bodies, such as rivers, lakes, or seas that have very low or no oxygen. Anoxic conditions in water lead to complex changes in biological and chemical processes and result in the production of gases such as hydrogen sulphide and methane. Dead zones are found along the shores of Lake Erie, the Black Sea and the Gulf of Mexico.

Lakes in an advanced state of eutrophication may be called hypertrophic, sometimes referred to as a 'dead lake', such as Lake Erie in the 1950s and 1960s. The Cuyahoga River, which drains into the lake, caught fire in 1969. Due to combustion of oil slicks, rubbish and other floating debris, this event combined with the condition of the lake, helped start the 1960s environmental movement.

4.4.9 Movement of Nutrients and Sediment in the Landscape

Water carries nutrients into the lake down any of the three main pathways: surface run-off, throughflow, or from the groundwater. Nitrates are water soluble and can follow any of these pathways. Phosphates are not very soluble and so tend to move attached to sediment or sewage. Mostly, phosphates will enter the lake from surface run-off or via sewage pollution. To understand and evaluate catchment management you need to consider how the flow of water is influenced by soil management, for example, compaction from machinery or livestock leading to reduced infiltration and more surface run-off.

Look back at Figure 4.2 to review the processes of water movement overland.

4.4.10 Applying the Pollution Management Model to Eutrophication

Level		Strategy
Cause	Agriculture	• Contour ploughing to reduce run-off • Reduce access of livestock to water courses • Reduce amount of fertiliser applied • Timing of fertiliser application to avoid rain
	Domestic	• Phosphate free washing powder • Dry composting toilets
Release and transfer	Agriculture	• Buffer zone around streams • Lighter tractors with broader wheels reduce soil compaction, so decrease surface run-off
	Sewage treatment	• Fertiliser production from sewage waste to harvest nutrients • Phosphate strippers (remove phosphate in sewage works)
Effects	Management of the ecosystem	• Pumping mud from lake bed • Removal of productivity as a usable resource (fish, algae for paper or fertiliser) • Use of decomposing batrley straw bales to inhibit algae • Biomanipulation, for example, removing fish to encourage invertebrate grazers

Table 4.12: **Summary of some strategies available to tackle eutrophication**

4.4.11 Evaluating Eutrophication Management Strategies

To evaluate a management strategy, consider how easy it would be to enact against the level of the strategy (cause, release and transfer or effect) and how effective it would be in controlling the pollution. Address the following points for a detailed evaluation:

- Difficulties of influencing human behaviour without legislation or economic incentive
- Catchment management agreements may be difficult to monitor
- Changes may be costly
- Disturbing habitats may have negative effects on biodiversity
- Consequences of biomanipulation may not be predictable
- Point source pollution control is easier to apply and monitor
- Phosphate is the limiting nutrient for nitrogen-fixing blue green algae
- Phosphate mainly transfers through detergent, sewage and surface run-off as it is non-soluble
- Nitrate dissolves readily and can follow all pathways in the hydrological cycle.
- Pumping mud may have other impact on the water quality (for example, increasing turbidity)
- Barley straw bales inhibit blue-green algae growth but do not prevent other algae.

4. WATER AND AQUATIC FOOD PRODUCTION SYSTEMS AND SOCIETIES

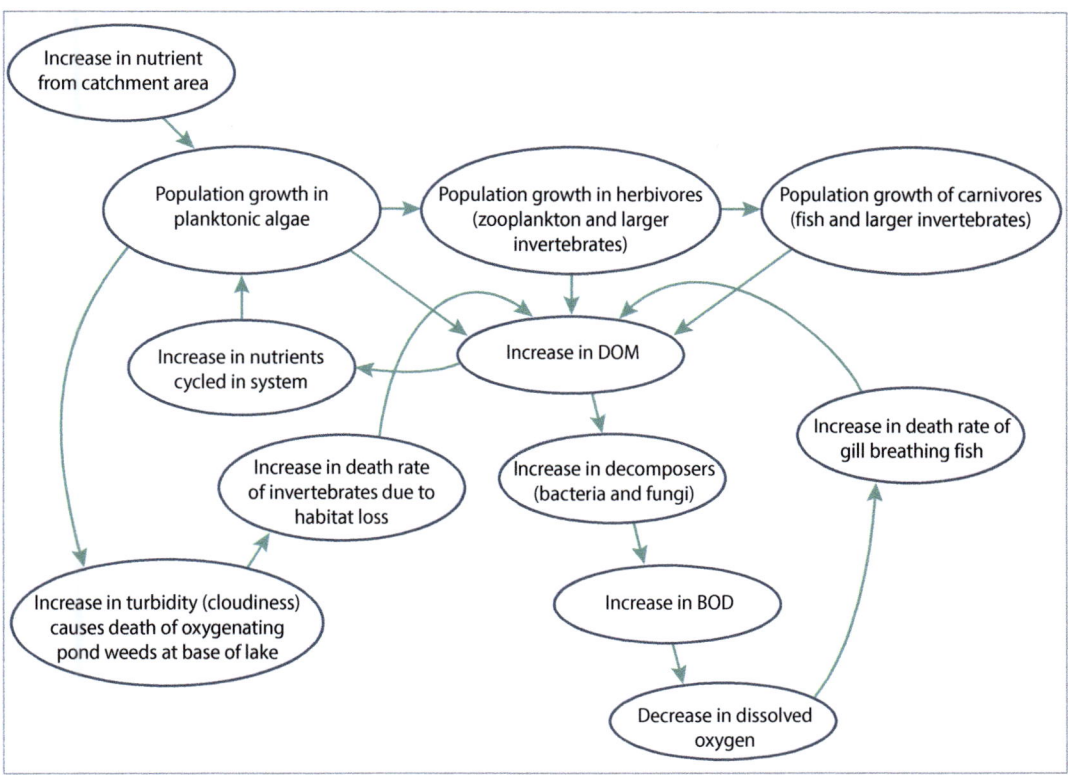

Figure 4.6: **Flow chart showing feedbacks that can occur during eutrophication**

Practice questions: Water Pollution

1. *State* what BOD stands for and *define* this term.

   ....

   ....

   ....

   ....

2. *Describe* how BOD is measured and used to indicate organic pollution in water.

   ....

   ....

   ....

   ....

   ....

ENVIRONMENTAL SYSTEMS AND SOCIETIES SL

3. Look at the case study data comparing two rivers sites (4.4.5). Analyse both the direct and indirect data.

   (a) **Explain** why the biotic index in the case study is lower for the second site.

   ....
   ....
   ....
   ....

   (b) **Explain** the results calculated for the Simpson's diversity index in the case study.

   ....
   ....
   ....
   ....

   (c) Evaluate the uses of direct measurement and indicator species in this study.

   ....
   ....
   ....
   ....

4. **Draw** a flow chart to describe the process of eutrophication and **outline** the process of eutrophication.

   ....
   ....
   ....

4. WATER AND AQUATIC FOOD PRODUCTION SYSTEMS AND SOCIETIES

5. *Identify* three positive feedbacks in eutrophication using Figure 4.6.

   ....

   ....

   ....

   ....

6. *Outline* some of the impact of eutrophication on named freshwater and marine environments.

   ....

   ....

   ....

   ....

7. Briefly *describe* pollution management strategies for eutrophication using the following table:

Level	Strategy	
Cause	Agriculture	
	Domestic	
Release and transfer	Agriculture	
	Sewage treatment	
Effects	Management of the ecosystem	

8. *Evaluate* the strategies at each level in the table above.

   ....

   ....

   ....

   ....

127

Chapter 5: Soil Systems and Terrestrial Food Production Systems and Societies

5.1 Introduction to the Soil System

5.1.1 Soils as Living Systems

Soils are the surface layers of weathered material from the Earth's crustal rocks, including organic materials and communities of living organisms. They are living ecosystems and a zone of interaction between the atmosphere, hydrosphere, lithosphere (rocks) and the biosphere (ecosystem above). Consider in the diagram below how there is a two-way cause and effect interaction between each of the components.

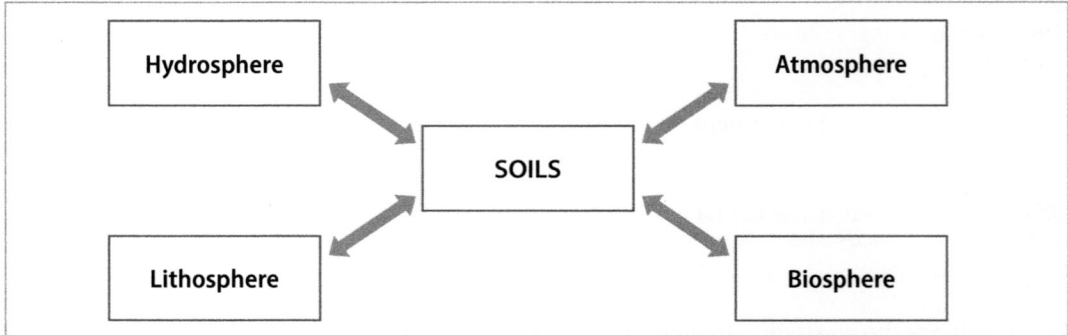

Figure 5.1: **A diagram showing interaction between soils and the major spheres of the earth**

This can be considered in terms of material interactions, including both transfer and transformation processes. The water cycle moves through the soil by infiltration and water may evaporate from the surface. The atmosphere may contain particulate matter that is deposited on the soil and particles may blow up into the atmosphere. Rocks in the lithosphere weather to form soils, and soils at depth and pressure may form rocks. Plants in the biosphere may extract nutrients from the soil and dead plants may end up forming parts of the soil. Consider this diagram in relation to the profile below.

5.1.2 Soil Profiles and Processes

Soils can be examined by digging down, normally a metre or so, to reveal the soil profile. This shows horizons (layers) that indicate changing composition with depth. Soil profiles show the transition from the living to the non-living components of the ecosystem. The kinds of soil profiles found vary according to climate, relief, and geology.

5. SOIL SYSTEMS AND TERRESTRIAL FOOD PRODUCTION SYSTEMS AND SOCIETIES

	Description and variations	Key processes
O Horizon	A dark layer containing organic material, when fully decomposed it is called humus. Deepest in waterlogged and cold conditions where decomposition rates are slow, such as tundra. Shallow or none where little organic matter is, e.g. desert, or where decomposition is fast, e.g. tropical rainforests.	Organic matter is added from the ecosystem within, e.g., leaf litter or dead organisms. Organic material is broken down into fragments by detritivores and decomposed by fungi and bacteria.
A Horizon	This layer of soil is the mixed layer, a mixture of organic and mineral particles. The mixed layers vary in depth depending on the living community in the soil. The deepest mixed layers are normally found under temperate grassland biomes due to the abundance of earthworms. In temperate deciduous forest the acid conditions reduce the number of earthworms and so the mixed layer is shallow.	Soil organisms mix organic material into the layer and help aeration, including earthworms, larger burrowing animals and plant roots. Leaching begins from the A horizon with water removing products of weathering and decomposition. Leaching can make the horizon both acidic and nutrient poor.
B Horizon	This layer consists of weathered mineral materials. The depth of the mineral layer varies greatly according to the climate and geology. Tropical soils have deep weathered layers often due to the chemical weathering of the underlying rocks. Leaching in soils occurs when rainwater dissolves useful nutrients in the soul and as the water infiltrates they are moved down into the upper part of these horizons.	Weathering and leacing are important processes here. Some minerals such as iron may builf up as a result of leaching and in leached soils the pH is more alkaline. In waterlogged soils low oxygen content can lead to few living organisms in this horizon.
C Horizon	This layer is made up of the underlying rock type, in various states of weathering.	Chemical and physical weathering of this material is of great importance in producing the soil type.

Figure 5.2: **Generalised soil profile**
Source: Adapted from 'Horizons' [CC BY-SA 4.0], via Wikimedia Commons/user:Wilsonbiggs

5.1.3 Soil Texture and Primary Production

Soil properties vary considerably according to the texture of the smaller particles in the soil, the components that are under 2 mm in diameter. These particles are produced by different weathering processes, of different minerals. They are broken down into three groups: clay, silt, and sand in increasing size order. These three particles may be present in different amounts in the soil as shown in the soil texture triangle (Figure 5.3). The varying combinations of these particles give soils their texture.

To read the soil texture triangle there are three axes that run parallel to the base of the triangle for each scale at zero. The soil at point **A** would have a mixture of approximately 35% clay, 12% sand, and 53% silt.

ENVIRONMENTAL SYSTEMS AND SOCIETIES SL

Point **A** is referred to in the text on page 129, and point **B** is a reference to practice question 2 on page 140

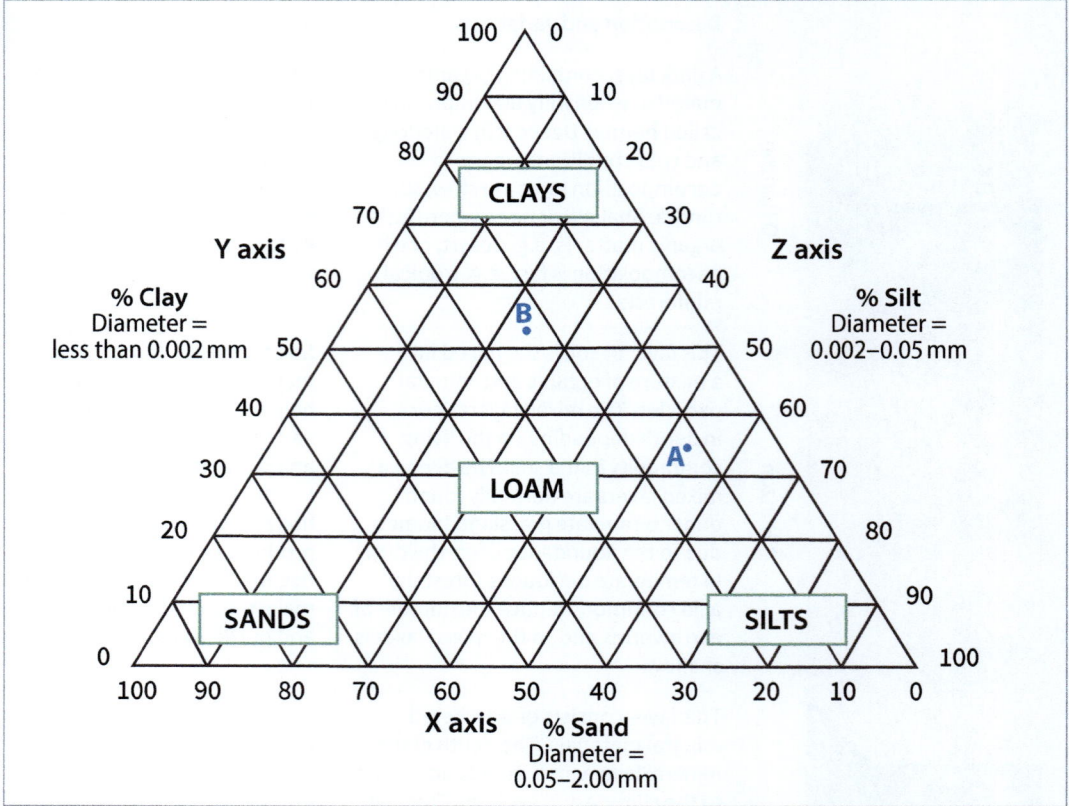

Figure 5.3: **Soil texture triangle**

Loams are mixed soils and, generally, the most fertile agricultural soils, because they retain nutrients and water, which are then accessible to plants. Loams also have the air spaces needed for plant root respiration. Clays hold onto the water and don't allow air to access the roots. Sands hold little water or nutrients.

	Water Retention and Availability	Nutrient Storage Capacity	Air Space	Primary Production
Clay	Sticky and easily waterlogged	High	Low	Medium to Low
Sand	Fast draining soils dry out easily	Low	High	Low
Loam	High to medium	Medium	Medium	High

Table 5.1: **Summary soil properties in clay, sand, and loam**

5.2 Terrestrial Food Production Systems and Food Choices

5.2.1 An Overview of Trends in Food Production and Soil Resources

Food production has grown considerably due to improvements in agriculture. Productivity of modern intensive agricultural systems has increased by a factor of at least 10 times per unit area and over 100 times per capita of labour.

In the More Economically Developed Countries (MEDCs), agriculture became more intensive during the 20th century, following the Second World War. We now have a far greater production of food and a far greater variety available to the consumer. Farmers have the help of improved machinery, which saves manual labour, agricultural chemicals

5. SOIL SYSTEMS AND TERRESTRIAL FOOD PRODUCTION SYSTEMS AND SOCIETIES

such as fertilisers and pesticides, and genetically modified crops that are pest resistant or grow faster or are resistant to pesticides. Farmers can now produce more food with fewer people. However, diversity of produce from such farms has declined.

In many Less Economically Developed Countries (LEDCs), the influence of a number of MEDCs was important, changing both systems of land ownership and methods of agricultural production, in particular, the development of high-yield seeds during the 'green revolution'. Farmers have been improving seed yields by artificial selection since the beginning of agriculture, but the term 'green revolution' came from tests in North West Mexico during the 1960s. Improved varieties of wheat increased yields massively; they were called 'miracle seeds'. However, increases in yield were also the response of the seeds to irrigation and fertilisers. The green revolution came at a cost as increases in fertilisers and pesticides are an essential part of keeping yields high. In many areas the yields that were boosted in the 1960s and 1970s began to fall. Where yields are not actually declining, the rate of growth is slowing rapidly or levelling off, as has now been documented in China, North Korea, Indonesia, Myanmar, the Philippines, Thailand, Pakistan, and Sri Lanka.

Improvements in food production are uneven. Modern techniques in farming have led to the production of surpluses in Western countries such as the 'beef mountains' and 'wine lakes' produced in Europe, starting in the 1980s and continuing today. At the same time, famines have resulted from both drought and increases in farming intensity, for example, in the Sahel of sub-Saharan Africa. The least developed countries are most vulnerable to starvation from famine because they lack the ability to import food in times of need. The export crops produced by such countries, such as cotton and groundnuts, are irrigated and improved; in comparison, subsistence agriculture is normally rain-fed and produces cassava, wheat and sorghum.

As human populations increase their demand for food through growth and development, there has been a push to use more marginal land in poorer countries and an increase in intensive farming in richer countries. Land with quality soil resources is in decline, and pressures that cause soil degradation increasing. Energy intensive food production systems, such as industrial scale greenhouses or plant factories, are becoming more common in richer countries.

5.2.2 Food Waste Issues in MEDCs and LEDCs

A large amount of food is wasted at the post-production stage, with the vast amount of this loss occurring in the production to retail stage of the food industry, significant amounts also arising in MEDCs as post-consumer waste. Post-consumer food waste from Europe and North America is similar to total food production in sub-Saharan Africa. The total amount of food waste has been estimated at around 1/3 of the food produced, enough to feed over 800 million people. The reasons for food waste vary considerably: in MEDCs the reasons are largely due to lifestyle, regulations and marketing; in LEDCs, due to lack of technology such as refrigeration or transport. See below:

MEDCs
- Retail food is wasted for not fitting market or regulatory standards.
- Domestic waste is thrown out as rubbish.

LEDCs
- Post-harvest loss occurs during handling and storage on farms.
- Processing and distribution loss occurs due to spoilage.

5.2.3 Cultural Context of Foods and Impact on Human Trophic Levels

Cultural context has considerable impact on what is seen as a food resource. Both religion and tradition inform what is acceptable to eat and what isn't. Many cultures in tropical regions eat insects from grasshoppers to moth and beetle larvae. Culture also influences what species of animal or parts of an animal are used, though this is also clearly linked to development. Poorer countries and societies use more of an entire animal's body. The cultural context may also influence the choice of societies to harvest food from higher trophic levels, for example, Japanese consumption of bluefin tuna, Mongolian mutton, horse and dairy diet, or American consumption of beef.

Food harvested from the lower trophic levels is generally more energy efficient per areas of land than higher trophic levels, though other considerations should be brought in regarding sustainability (see section 5.2.5).

Studies of human trophic levels (HTL), which is the average trophic level that humans are feeding at, reveal that it is generally increasing but that these trends are not the same everywhere in the world. The research indicates that there are five main groups; note that these trends are also linked to development and urbanisation.

Group	Countries and Regions	Diet
Group 1: Low and stable HTL	South East Asia and sub-Saharan Africa	Mostly plant-based; little change in diet noted
Group 2: Low and increasing HTL	China, India and others in Asia, Africa, and South America	Meat intake increasing; linked to increasing urbanisation and development
Group 3: Higher HTLs than Group 2 and increasing	Central America, Brazil, Chile, Japan, and Southern Europe	
Group 4: High HTL and decreasing	US, Northern and Eastern Europe, Australia, and New Zealand	Decline in meat intake, due to environmental, ethical, and health concerns
Group 5: Highest HTL and decreasing	Mongolia, Scandinavia, Mauritania, and Iceland	Decline in meat intake; the only group that includes countries with rising GDP and urbanisation that is decreasing HTL

Table 5.2: **Changes in the human trophic level**
Source: Based on Bonhommeau, et al., 2013

5.2.4 Choice of Food Production Systems

Choice of food production systems can be seen to be influenced by a variety of factors: socio-economic, cultural, ecological, and political. The reasons behind such choices are complex, but here are two examples to illustrate how these factors work together. These examples from the USA and South East Asia are further developed in section 5.2.5.

South East Asian hill tribes vary in number and typically have poorer economies, low population densities, and traditional nomadic life styles that allow for subsistence agricultural practices such as shifting cultivation of upland glutinous rice and mixed cropping. Nomadic lifestyle is rarer due to pressure on the land, but the traditions are still present. Animist beliefs are common, with ceremonies linked to seasonal cycles and agriculture practice. In an Akha hill tribe village, a new-born child is considered impure until the 13th day, then rice is rubbed on the child's lips as the first rice meal. The first grains of rice planted each year are washed in purified water, and a spirit house is built in the field where nine grains are planted. Later these plants will be harvested first and

5. SOIL SYSTEMS AND TERRESTRIAL FOOD PRODUCTION SYSTEMS AND SOCIETIES

offered to the ancestors (Anderson, 1993). Use of rice and forest products is tightly bound to culture and tradition from birth to death in these communities.

Richer societies with higher population densities and technology lead to establishing permanent fields, irrigation and use of machinery and fertiliser to maintain productivity. Commercial cereal farming in the US shares its roots with the birth of capitalism in rural Europe. Enclosure of land to claim legal ownership marked the beginning of the capitalist approach. Ownership of land passed through a period of family farm ownership as seen in much of Europe and America until the second half of the 20th Century (Newby, 1987). At the time of the Dust Bowl events, farms were mostly family farms, and after the Second World War the growth of agribusiness saw many of these farms amalgamated and diversity of production lost to monoculture. This is often linked to the requirements of supermarkets for a regular supplier of a single, standardised product. Modern agricultural chemical companies own many aspects of food production systems, including genetically modified seeds that are resistant to the company's herbicide products. Production of food in MEDCs is often now tightly bound to market requirements, perceived consumer preference, and contract requirements to deliver food to large-scale supermarket economies.

5.2.5 Comparison of Two Food Production Systems

	Energy output (10^9 J/ha/yr)	Energy Input (10^9 J/ha/yr)	Energy Ratio (Energy Out/Energy In)
Case study 1: subsistence agriculture in South East Asia	15	0.5	30
Case study 2: commercial cereal farming in America	50	50	1

Table 5.3: **A comparison of energy ratios in agricultural systems**
Source: Includes data from Pimental, 2009; and Fluck, 1992

These two systems are different in terms of the degree of openness, in other words, how much the system requires inputs and outputs from far away. Subsistence farming uses the available resources supplied from the local area as input, as output the food is used to supply local needs. The commercial system involves more import of resources from further away, with machinery, fertiliser, and labour brought in to the area. Outputs supply a national or global demand through distribution systems operated by supermarkets and using long distance haulage.

In subsistence farming the land is left fallow and nutrients regenerate due to succession. Mixed cropping is commonly practised with a wide diversity of crops such as banana, papaya, peach, kidney bean, soya bean, peanuts, and sesame. Gathering of secondary forest produce such as mushrooms, honey, bamboo shoots, and ant eggs is also significant in the local diet. Traditional management does not involve piped extraction from streams, or other forms of irrigation, but works with the seasonal cycles of the dry and rainy season.

Judged against the indicators of sustainable agriculture (see section 5.2.6), subsistence farming appears to be the more sustainable. However, only small numbers of people can be sustained per unit area from this kind of agriculture given the limited total output. If this was more widely practised then greater areas of land would be needed to meet demand for food, and supply of food to urban areas would become difficult. The impact on destruction of habitats and conservation of biodiversity would be considerable.

5.2.6 Sustainability of Terrestrial Agriculture

Sustainable agriculture can be defined as that which:

- Produces good quality food efficiently and inexpensively
- Minimises resource use
- Protects soil, water and air quality
- Preserves and enhances biodiversity and landscape quality.

Energy efficiencies vary considerably; stock can be compared in terms of the food conversion ratio, a measure of the animal's efficiency in converting feed to body mass. For example, Table 5.4 shows that 2.2 kg of food is used to produce 1 kg of chicken, compared with 12.7 kg for 1 kg of beef. Compare these with the figures for aquatic systems in Table 4.8 on page 116.

Table 5.4: **Feed conversion ratios for terrestrial systems by animal type**
Source: Adapted from FAO, 2013

Animal	Efficiency Ratio (kg of feed/kg of total live weight)
Insects	1.5
Poultry (Meat)	2.2
Pig	5.8
Beef	12.7

Any food production system of plants will have a better energy yield, per area of land. But there may be other environmental impacts to consider. For example, cauliflower farming on coastal hills causes soil erosion and coastal eutrophication, whereas sheep farming on the same land suffers less soil loss.

In commercial farming the high output is sustained by high input and much of this is dependent on using non-renewable resources. Improvements are often fossil fuel dependent as shown in Table 5.5.

Irrigation	To pump water
Mechanisation	Fuel to run machines
Preservation	Electricity to refrigerate
Fertilisers	Dependent on electricity for production
Distribution	Distribution of input and outputs require fuel for transport

Table 5.5: **Fossil fuel requirements in modern agriculture**

Intensive farming is practised mostly as monoculture, with famers focusing on an output of a single crop such as corn. Commercial agricultural systems rarely use the natural regeneration capacity of the soils as this process is slow and requires land to be put into fallow.

5. SOIL SYSTEMS AND TERRESTRIAL FOOD PRODUCTION SYSTEMS AND SOCIETIES

5.2.7 Improving Sustainability in Terrestrial Agriculture

More sustainable agriculture will be developed if societies demand it. There are a number of approaches towards this from various sectors of society as outlined in Table 5.6.

Method	Focus	Benefits to Sustainability
Altering human activity	Reduction of meat in diet, in particular, least efficient livestock, e.g., beef	Increases total carrying capacity for humans Health benefits to consumer
	Reduction of food miles Promotion of local produce, e.g., farmer's markets	Reduces additional environmental impact from fossil fuel combustion Reduces food waste from transport
	Promotion of organic foods	Reduces impact on the environment from pesticide bioaccumulation.
	Food labels to assist consumer choice, e.g., soil association standards or equivalent	Allows consumers access to relevant information regarding environmental, social, and health impacts of food
Monitoring and control of standard effects	Governmental and intergovernmental organisations	Regulation and setting of stringent standards of food production for corporations
	NGOs	Raises consumer awareness on all aspects Lobbies governments to improve regulations

Table 5.6: **Roles of different social sectors for improving agricultural sustainability**

5.3 Soil Degradation and Conservation

5.3.1 Soil Development and Succession

Soils can take a very long time to develop; the processes described in section 5.1 may take hundreds of years to result in fertile soils through succession of both the soil community and ecosystems or agricultural system above it. Modern crop agriculture tends to side step this process; instead of relying on natural process to restore soil fertility it relies on artificial inputs of fertiliser. Mature soils with intact communities are more resistant to most forms of degradation.

5.3.2 Soil Degradation Processes and Consequences

Soils are degraded by a variety of human activities as summarised in Table 5.7. In general, a degraded soil has lost physical, chemical, or biological structures that give it fertility. Soils may also become toxic due to pollution.

	Processes	Consequences
Overgrazing	Grazing by livestock at high density damages physical structures by trampling and can remove vegetation cover.	Soil erosion is increased by wind or water. Compaction of soil occurs (increase in density). Overgrazing can lead to desertification.
Deforestation	Removal of forest cover removes protection by canopy and roots.	Soil erosion increases.
Unsustainable agriculture	Most agricultural practices remove the upper soil horizons (O and A), completely planting in the B horizon.	Agricultural systems become dependent on fertiliser application.
Irrigation	Poor irrigation techniques can quickly lead to soil salinisation from evaporating water. Salts may come from the soil or from salt deposits in deep wells.	Salinisation causes a stress for plants and can damage agricultural productivity or lead to crop failure.
Erosion	Wind and water can remove upper layers of the soil, removing organic material, minerals, and nutrients.	Water erosion impacts on water quality and can cause flooding. Wind erosion impacts on air quality (dust).
Desertification	Enlargement of deserts is often connected to soil degradation. It occurs in drier climates, often along the edge of deserts. Has been predicted in rainforest as a consequence of climate change.	It can cause crop failure and lead to malnutrition and famine.
Urbanisation and Industry	Soil is directly removed for other land use and replaced with hard wearing surfaces like concrete. Pollution occurs to surrounding soils from contaminated run-off.	Soils can become toxic due to heavy metals or other poisonous chemicals.

Table 5.7: **Soil degradation processes and consequences**

> **Weathering** is the breakdown of parent material (rocks) and helps to form soil.
> **Erosion** is the movement of weathered products, including soil, by wind, water and slope processes.

5. SOIL SYSTEMS AND TERRESTRIAL FOOD PRODUCTION SYSTEMS AND SOCIETIES

5.3.3 Soil Conservation Measures

Strategies to conserve soil are designed to reduce the processes of soil degradation described above. They are outlined in Table 5.8.

Strategy		Description
Cultivation techniques	Terracing	Terraces reduce surface run-off and erosion. It is labour intensive and requires co-operation between land owners.
	Contour ploughing	Ploughing along the contour lines, instead of up and down the slope, reduces surface run-off and erosion.
	Zero tillage	Soil is not ploughed but seeds are dibbled into the ground with minimal disturbance.
Wind reduction	Wind breaks or shelter belts	Lines of trees, shrubs, or hedges are designed to prevent wind erosion.
	Strip cultivation	Mixed cropping in a systematic series of bands acts as a barrier to water or wind. Staggered harvest reduces the amount of exposed soil.
Soil conditioning	Liming	Lime is added to the soil to reduce acidity. Soil acidity can be natural or caused by acid deposition (see section 6.4).
	Mulching	Covering the soil with organic or inorganic material can protect it from erosion, prevent weed growth, and restore nutrient levels. Straw or chippings can be used and also waste materials, e.g., old carpets and newspapers.
Improved irrigation	Trickle drip	Slow release of water from pipes under the surface can help reduce loss of water by evaporation and salinisation.

Table 5.8: **Soil conservation measures**

5.3.4 Case Study to Evaluate Soil Management Strategies

You need to evaluate soil management strategies in two systems, one a commercial farming system and the other a subsistence farming system. In your evaluation consider how well the management is reducing soil degradation.

Northern Thai hill tribe subsistence farming

Monsoon rainfall can lead to rapid rates of erosion on bare ground and the detail of how soil is managed effect rates of erosion. The societies of the Northern Thai hills have diverse ethnic backgrounds and a variety of traditional strategies towards subsistence agriculture. Mostly they practise a form of rotational slash and burn clearance. This involves clearing forests through cutting and burning, and, after cultivation, returning the fields to a fallow stage in which the land is not used for planting. Table 5.9 summarises the traditional agriculture of Karen and Mien (Hmong) societies as recorded during the 1970s (Kundstadter, Chapman, & Sabhasri, 1978).

Society	Elevation	Agriculture	Major Crop	Cultivation	Soil Fertility
Karen	Lowland (500–1000 m)	Short cultivation with long fallow	Rice	Cleared and weeded by hand tools	Stable due to fallow period; villages permanent
Mien	Uplands (1000 m+)	Long cultivation Very long fallow	Corn (maize) and opium (cash crop)	Deep hoeing of soil	Long-term decline; nomadic lifestyle

Table 5.9: **Summary of traditional soil management by South East Asian hill tribes from the 1970s**

Increases in population density have led to changes in agriculture and lifestyle for the hill tribes. Nomadic lifestyles and opium growing have mostly ended; this combined with a general increase in commercial cropping means that the approaches described have now been replaced by permanent agriculture improved by using irrigation and fertiliser. Many alternative commercial crops such as coffee, cabbages, soybean, ginger and peanuts, are now grown. Farmers and environmental researchers recognise that steeper slopes are the least desired for agriculture and lead to the highest rates of erosion. Rates of erosion on steep slopes have been recorded at 64 tonnes per hectare, compared with averages of 24 tonnes per hectare for the shallower slopes (Forsyth & Walker, 2008).

Commercial cereal farming in United States of America

The southern part of America's Great Plains has natural vegetation that is dry prairie grassland, bordering on an extensive desert. This region was the site of an environmental catastrophe in the 1930s called the 'Dust Bowl', when intense wind erosion filled the skies with dust from the farmer's fields. Severe drought caused by a change in weather patterns combined with rapid intensification of agriculture. Air quality suffered as topsoil was stripped forming 'black blizzards' and visibility in some areas of fell to under a metre. Soil was transported over a thousand kilometres and fell like snow in Chicago and even as a red rain in New England.

Decade	Agriculture	Major Crops	Cultivation Technique	Effects on Soil Fertility
1930s	Extensive ranching and cotton farming	Beef Wheat Corn Cotton	Extensive herding of cows Deep ploughing using a steel plough Winter fallow and stubble burning	Overgrazing led to loss of protective vegetation cover
2000s	Intensive farming of cereal crops	Beef Wheat Corn Soybean	Summer fallow Intensive irrigation	Improved by use of better farming but also fossil fuels and irrigation

Table 5.10: **Summary of soil management changes in the Great Plains**

Soil conservation techniques were enforced through the soil conservation act of 1935. This included the promotion of strip farming and shelter belts. The development of the heavy-duty chisel plough, and changes in agricultural practice such as summer fallow, helped to further protect the soil. The farming system links to the management of water resource in the Ogallala aquifer, as outlined in section 5.3.5.

5. SOIL SYSTEMS AND TERRESTRIAL FOOD PRODUCTION SYSTEMS AND SOCIETIES

5.3.5 Case Study of the Ogallala Aquifer and Its Use in Farming

The Ogallala or high plains aquifer stretches from South Dakota to Texas and includes the dust bowl states of Oklahoma and Kansas. This aquifer supplies much of the water requirements to the agriculture of the Great Plains described previously. The importance of irrigation can be seen clearly in this aerial image of the ground showing irrigated crop patches in neat circles from a central pivot system of irrigation. The larger irrigated circles are over 1,500 m in diameter.

This aquifer is one of the world's largest and provides water for over 25% of irrigated land in the United States. Annual rates of extraction are over 25 km^3/year. The water in the aquifer is sometimes termed 'fossil' water, as much of it was supplied into the current store at the end of the last ice age. The aquifer occupies sedimentary rocks and at the thickest section is around 300 m in depth. Estimates of the total volume suggest that the aquifer has already depleted by around 10% since extraction began in the 1950s. Water levels have dropped by over 75 m in areas of intense irrigation like Oklahoma.

The aquifer now recharges slowly through rainfall, snow melt in the Rocky Mountains, and through seasonal wetland ecosystems called prairie potholes or playa lakes. These small temporary surface water features are disappearing due to urbanisation and agricultural extraction before they can recharge the aquifer. Rates of extraction are so much greater than recharge rates (only a few cm a year) that the extraction is often referred to as water mining or overdraft. Climate change is expected to further reduce the ability of the aquifer to recharge. The aquifer continues to decline at rates of over 80 cm a year on average and estimates give the aquifer a future life span of only 25 years.

Figure 5.4: **Aerial image of central pivot irrigation systems in Oklahoma**

Source: NASA/METI/AIST/Japan Space Systems, and US/Japan ASTER Science Team

ENVIRONMENTAL SYSTEMS AND SOCIETIES SL

Practice questions: Soil Systems and Terrestrial Food Production Systems and Societies

1. **Outline** how soil systems integrate aspects of living systems.

2. **Estimate** the proportions of sand, silt and clay in the soil marked as **B** in Figure 5.3 on page 130.

3. **Compare and contrast** the properties of sand, clay, and loam, including the effect on primary productivity.

Compare and contrast
Answers must include sentences that compare effectively. The best answers will use comparative clauses, adjectives, or verbs. Tables can be used if they make effective comparisons linked clearly in a row.

4. **Outline** the processes and consequences of soil degradation.

5. **Evaluate** soil management strategies in a named subsistence and a named commercial farming system.

5. SOIL SYSTEMS AND TERRESTRIAL FOOD PRODUCTION SYSTEMS AND SOCIETIES

6. **Outline** the evidence that the global food supply system is imbalanced.

　

7. Consider the following quote from the ecologist Howard Odum in 1971.

   > *A whole generation of citizens thought that the carrying capacity of the earth was proportional to the amount of land under cultivation and that higher efficiencies in using the energy of the sun had arrived. This is a sad hoax, for industrial man no longer eats potatoes made from solar energy, now he eats potatoes partly made of oil.*
   >
   > Cited in Ramage & Shipp, 2009

   As we reach peak oil production, **discuss** this statement in relation to food supplies.

　

8. **Compare and contrast** the efficiency of terrestrial and aquatic food production systems.

　

9. **Compare and contrast** two food production systems in terms of inputs and outputs of materials and energy and the system characteristics.

10. ***Describe*** the environmental impacts of the two systems you have compared.

   ....
   ....
   ....
   ....

11. ***Evaluate*** the sustainability of two contrasting terrestrial food production systems.

   ....
   ....
   ....
   ....

12. ***Discuss*** the links that exist between social systems and food production systems.

   ....
   ....
   ....
   ....

Chapter 6: Atmospheric Systems and Societies

6.1 Structure and Composition of the Atmosphere

6.1.1 The Atmospheric System

The gases that make up our atmosphere are mostly concentrated in a 10 km layer above the earth's surface. These gases are in a dynamic state, they are not constant or consistent over the Earth, or through the atmospheric layers, and have changed over time. The oxygen in the atmosphere was produced by the photosynthesis of blue-green algae in the early stages of the Earth's evolution around 2.8 billion years ago. The mixtures in the air we breathe are in a state of dynamic equilibrium. Inputs and outputs to the atmospheric system by chemical and biological processes have changed throughout the history of life on Earth. In recent years the human impact on the atmosphere has become greater in scale and more significant on its ability to support life on Earth. Changes in ozone, carbon dioxide, and water vapour all have significant impact on ecosystems and people.

6.1.2 Composition of the Atmosphere

The atmosphere is made of air, a mixture of gases. Two gases dominate the mixture, nearly 80% is nitrogen and most of the rest is oxygen.

Gas	Percentage (%)
Nitrogen	78.1
Oxygen	20.9
Argon	0.9
Carbon dioxide	0.04
Water	Variable
Trace Gases (e.g., methane, ozone, and hydrogen)	All under 0.002

Table 6.1: **Composition of the lower atmosphere**
Source: Includes data from Wright & Nebel, 2002

Nitrogen is a stable component of the atmosphere. Globally, carbon dioxide and oxygen levels fluctuate due to photosynthesis, respiration, decomposition and combustion.

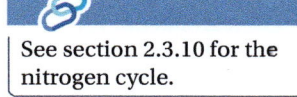

See section 2.3.10 for the nitrogen cycle.

Changes to the levels of carbon dioxide are much greater as a percentage amount as there is so little there to start with. Water is present in the atmosphere as a gas, water vapour, or as droplets of liquids in clouds.

6.1.3 Structure of the Atmosphere

There are two principal layers in the atmosphere. The troposphere is the lower layer from the Earth's surface up to about 10 km altitude; it is a turbulent layer containing most of the world's weather and the circulation described below. The stratosphere is much higher, around 10 km to 50 km, and forms a stable layer over the troposphere. There is some mixing between the two but generally they are divided by a layer called the tropopause. Sunlight travels through the troposphere and warms the surface of the planet. The hot surface of the planet warms the air from the base and it rises in convection cycles, later cooling and falling. Air temperature falls with altitude. In the stratosphere, heating takes place so that temperature increases with altitude. This is due to the action of UV light with ozone.

Altitude:
Height above ground level is altitude; height of ground above sea level is elevation.

6.1.4 Cloud Albedo

Clouds form in the atmosphere at many different levels within the troposphere. The highest clouds form up to 6–12 km high, with the tops at the border between the troposphere and stratosphere. The higher clouds are often formed of ice crystals, but otherwise clouds are formed from liquid water. They stop sunlight entering further into the atmosphere and reflect the light back to space, a process called 'cloud albedo'. Consequently, they can cool conditions on the Earth's surface below.

Albedo is also discussed in 1.3.9 and 1.3.11.

6.1.5 The Greenhouse Effect

Certain gases allow solar radiation in to the atmosphere, but absorb heat radiated back from the Earth's surface. This keeps the planet's surface warmer than it would be without the effect. This is explored in depth in section 7.2 on page 164.

6.2 Stratospheric Ozone Depletion

6.2.1 The Stratospheric Ozone Layer

Ozone (O_3) is triatomic oxygen. Oxygen can exist in three forms: monoatomic (O), diatomic (O_2) and triatomic (O_3). Ozone is found naturally in the stratosphere from 15 km up to over 60 km. Natural concentrations of ozone are at a very low density of around 12 molecules per million air molecules.

Ozone is formed in a series of chemical reactions that occur when ultra violet light collides with oxygen. Oxygen is split from diatomic to monoatomic forms (I), which react in a series of stages with other chemicals. Ozone can form from a combination of monoatomic and diatomic oxygen (II). Ozone can then split again by reacting with UV forming more reactive monoatomic oxygen (III).

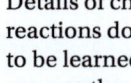

Chemical equations
Details of chemical reactions do not ***need*** to be learned, but you can use them if you can remember them!

I.	$O_2 + UV \rightarrow O + O$
II.	$O + O_2 \rightarrow O_3$
III.	$O_3 + UV \rightarrow O_2 + O$

These reactions cycle continuously in a dynamic equilibrium. Consequently, the formation and destruction of ozone absorbs the energy in UV light, preventing it arriving at the earth and warming the stratosphere.

6. ATMOSPHERIC SYSTEMS AND SOCIETIES

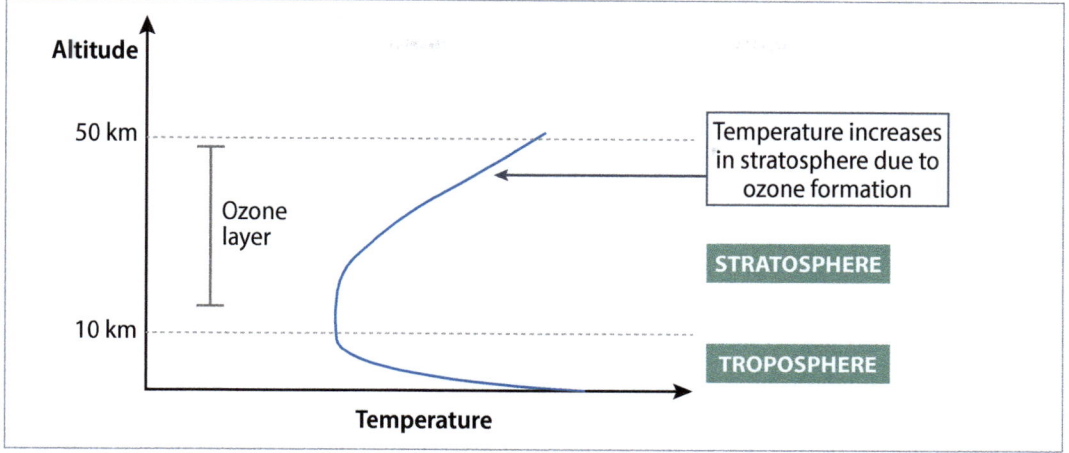

Figure 6.1: **Vertical structure of the atmosphere showing altitude of ozone layer**

6.2.2 Biological Significance of the Ozone Layer and the Effects of UV Light

The ozone layer is a product of the evolution of an oxygen-rich atmosphere as described above. It is of biological significance as it filters out UV light, particularly the form called UVb, which has a negative impact on many species that have evolved under ozone protection. The effects of excess UVb light include the following:

- Increases mutation rates in DNA causing cancers such as skin cancers in humans
- Can cause eye cataracts
- Can damage the ability to carry out photosynthesis in plants and phytoplankton
- Reduces primary production and, therefore, total productivity.

6.2.3 Depletion of Stratospheric Ozone

Ozone depleting substances (ODSs) include many manufactured chemicals such as halogenated organic gases including chloroflurocarbons (CFCs) and hydrochloroflurocarbons (HCFCs). They are or, have been, used as propellants in aerosols, coolants for refrigerators, and air conditioning, fire extinguishers, and blowing agents for packaging. Methyl bromide is an ODS that has been used extensively as a pesticide. Such chemicals are highly persistent in the environment. Due to their lack of reactivity, they move through the troposphere to the stratosphere, a process that takes decades. On arrival in the stratosphere, ODSs react with UV light, releasing the halogens they carry such as chlorine (Cl) and bromine (Br) (I). Halogens released react with ozone (II) and monoatomic oxygen (III).

> **Ozone hole and global warming**
> Avoid confusing these two terms and avoid writing them in the same sentence. A common error is to link them together wrongly.

I.	ODS + UV → product + Cl
II.	$O_3 + Cl \rightarrow O_2 + ClO$
III.	$ClO + O \rightarrow O_2 + Cl$

One single atom of chlorine may cycle through these reactions 100,000 times, disrupting the normal equilibrium of ozone formation and destruction. A steady decline in total ozone in the stratosphere of around 4% per year has occurred since the late 1970s. Ozone destruction has been most severe in the polar regions even though most of the ODSs are released in developed countries. ODSs carried up to the stratosphere through atmospheric circulation migrate to the poles and concentrate in a stable spring vortex of very cold air at temperatures of around −80°C.

The first confirmation of large scale ozone depletion was found over the Antarctic region in 1985 and became known as the 'Ozone Hole'. In March 2011 record depletion was observed over the Arctic. Ozone depletion events will continue to occur for some years to come, due to the lag time involved between emission and arrival in the stratosphere. The following timeline summarises some of the key events in the history of ozone depletion and its management.

OZONE DEPLETION AND ACTION

1928	CFCs discovered as highly versatile, non-toxic, non-corrosive, long-life gases
1957	Lovelock invents the electron capture detector and uses it to monitor CFCs
1974	Scientists Rowland and Molina hypothesize ozone depletion; findings opposed by industries such as the manufacturers of aerosols
1977	UNEP committee coordinates study of the ozone layer
1978	CFCs banned in the USA, Canada, Norway, and Sweden
1982	CFC consumption increases as industry demands proof of a connection
1985	**British Antarctic Survey discovers Antarctic ozone hole**
	Vienna Convention for the Protection of the Ozone Layer
1987	**Montreal Protocol on substances that deplete the ozone layer**
1990	London amendment to assist developing countries in reducing ODS production and use. Methyl bromide to be phased out
1996	CFCs no longer legally produced or consumed in many industrialised countries
1997	Montreal amendment on methyl bromide, licensing of import and export introduced
	Molina states Montreal Protocol is starting to work, based on halogen concentrations in the stratosphere
2000	**General trend since 2000 is shrinkage of the Antarctic ozone hole**
2010	**80% reduction of ODSs achieved since 1988**
2016	Kigali amendment to the protocol added to support action on climate change as some replacement chemicals were also powerful greenhouse gases

6.2.4 Management Strategies for the Ozone Layer

Applying the pollution management model from Table 1.10 to the ozone layer, a number of different strategies can be seen, examples of which are given in Table 6.2.

6. ATMOSPHERIC SYSTEMS AND SOCIETIES

Level	Strategy	
Cause	Refrigerants	Replace ODS dependent refrigeration with existing 'greenfreeze' technology driven by propane or butane
	Propellants	Pump action sprays instead of aerosols Alternative propellants to ODSs Alternatives to aerosols, e.g., shaving soap instead of spray foam
	Gas blown plastics	Alternatives to CFCs Alternative packaging
	Methyl bromide	Alternative pesticides Organic farming
Release and transfer	Refrigerants	Recycle old CFC coolants from old fridges and air conditioners
Effects	Protection from excess UVb	Sunglasses can protect excess UV from entering the eyes Sun block can protect from skin cancers

Table 6.2: **Management strategies for the ozone layer**

6.2.5 Evaluating Ozone Management Strategies and the Role of National and International Organisations

As a global pollution problem, ozone depletion required coordinated international action. This came through the collaboration of the scientific community and governments with the UNEP as the coordinating body (see the timeline summary). The chemical industry complied with the regulations and developed alternatives. The Montreal Protocol has allowed for unprecedented economic and political cooperation and is seen as success story by many environmentalists.

Public pressure also helped to effect change and successful campaigns from environmental NGOs such as Friends of the Earth helped to raise awareness during the 1980s in the UK. Greenpeace campaigns in 1996 helped demonstrate that 'Greenfreeze' technologies are viable options for commercial usage, although there has been reluctance to apply these on a large scale.

When evaluating the success of the actions, consider the following points:

- Illegal trade in CFCs was uncovered by the Environmental Investigation Agency (an NGO).
- Chemical companies are more interested in controlling the market for replacement gases than the impact of these chemicals.
- The annexes to the protocols list many gases that are replacements for previously listed gases.
- HCFCs replaced CFCs and were later discovered to be ODSs (but with a lesser impact).
- HCFCs were replaced by hydrofluorocarbons (HFCs) which are powerful greenhouse gases.
- Some impact on ozone can occur from natural events such as volcanic eruptions.
- Non-aerosol alternatives can be used but are not promoted as much as alternative chemicals, for example, traditional shaving soap instead of aerosol-propelled shaving foam or pump action sprays.

CFC, HCFC, and HFC
When revising, all of these chemicals names are ozone depleting substances. They have been brought in to replace one after another under the Montreal Protocol, Yet these replacements cause other problems and HFCs are also powerful greenhouse gases as well. They are now being phased out for this reason.

Practice questions: Structure and Composition of the Atmosphere / Stratospheric Ozone Depletion

1. *State* the names of the five most common gases that make up the atmosphere at the Earth's surface.

2. *Draw* a diagram to show the vertical structure of the atmosphere. Label the troposphere, stratosphere, and ozone layer. Indicate temperature change and circulation patterns.

3. *Explain* how UV absorption in the stratosphere is in steady-state equilibrium.

4. *Describe* how the arrival of halogenated gases in the stratosphere leads to ozone depletion.

5. State the ways in which UVb radiation affects living tissue and biological productivity.

6. ATMOSPHERIC SYSTEMS AND SOCIETIES

6. Complete the following table to show strategies to reduce stratospheric ozone depletion.

Level	Strategy	
Cause	Refrigerants	
	Propellants	
	Gas blown plastics	
	Methyl bromide	
Release and transfer	Refrigerants	
Effects	Protection from excess UVb	

7. Underline three methods of reducing the manufacture and release of ODSs in the table.

....
....
....
....

8. *Describe* and *evaluate* the role of national and international organisations in reducing ODS emissions.

....
....
....
....

149

6.3 Photochemical Smog

6.3.1 Primary and Secondary Pollutants from Urban Areas

Urban areas in both developed and developing countries suffer from localised air pollution problems. Urban air pollution is generated by transport, cooking food, dust from construction sites and roads, heating in homes, industry, and power generation. The main primary pollutants are:

- Particulate matter (much is carbon, but also other chemicals)
- Black carbon (soot)
- Sulphur dioxide
- Nitrogen oxides
- Volatile organic compounds (some toxic and carcinogenic)
- Carbon monoxide
- Lead.

Several secondary pollutants are also produced by reactions of these primary pollutants with each other and with sunlight. These secondary pollutants include tropospheric ozone and the mixture of gases found in a photochemical smog.

6.3.2 Tropospheric Ozone

Ozone is good in the stratosphere for life but bad in the troposphere in contact with living things. Ozone can cause lung damage, irritate the eyes, damage plants, and corrode fabric. It is a powerful oxidant and reacts corrosively with many materials.

I.	NO_2 + sunlight → NO + O
II.	$O + O_2 → O_3$
III.	$O_3 + NO → O_2 + NO_2$

At ground level, ozone is formed as a secondary pollutant by reactions involving the nitrogen oxides (NO_x). NO_xs are formed from reactions of atmospheric oxygen and nitrogen in internal combustion engines, most commonly cars. In particular, nitrogen monoxide (NO) and nitrogen dioxide (NO_2) are formed. NO_2 reacts with sunlight to form NO and monoatomic oxygen (I). This monoatomic oxygen reacts with O_2 to form O_3 (II), ozone. O_3 can react with NO to form NO_2 (III).

6.3.3 Photochemical Smog

Smogs are products of weather combined with pollution. Photochemical smogs require sunny, windless conditions so that primary pollutants are not dispersed and sunlight drives the reactions. Urban areas are often hotter due to lack of vegetation and water and the ability of concrete to absorb heat. This urban heat island effect may help to reduce dispersal and set up temperature inversions. Photochemical smogs are linked to higher population densities, particularly in low lying cities with higher mountains around, for example, Mexico City.

Dispersal of primary pollutants may be prevented by temperature inversions. These conditions develop in low lying areas particularly after cool clear nights which allow the ground to lose its heat. Firstly, the ground surface cools during the night and a layer of cool air sits on the Earth's surface. Cool air sinks and pollution entering the air in the early morning can prevent the sunlight warming the ground. Instead, the layer above warms and rises away from the cool, polluted air which stays trapped on the ground.

Tropospheric ozone is a major component of photochemical smogs, but these also include hundreds of primary and secondary pollutants, formed under the influence of sunlight. Hydrocarbons from car exhausts may enter reaction series to produce more dangerous chemicals, such as methane which reacts to form the strong oxidant peroxyacetylnitrate. Toxic and carcinogenic chemicals may also be present.

6.3.4 Deforestation and Economic Impact

In some areas deforestation may add significantly to the production of smogs, for example, in Singapore or mainland South East Asia where forest fires regularly pollute the cities in the burning season. Other sources of burning include food preparation from charcoal, or burning of waste materials. The economic impacts are large, with airports closed and people recommended to stay indoors due to health implications.

6.3.5 Management Strategies for Photochemical Smogs

Applying the pollution management model from Table 1.10 to the urban air pollution, a number of different strategies can be seen.

Level	Strategy	
Cause	Cars, buses, and taxis	Reduce demand for private cars through public transport Promote cycle and bus lanes Restrictions and tolls for car entry to urban areas Promote cleaner fuels and hybrid or electrical models
	Electricity	Reduce consumption of electricity through building design Small-scale green power on city buildings, e.g., solar, wind Locate power stations away from urban areas
Release and transfer	Cars, buses, and taxis	Catalytic converters help reduce NOx emissions Monitor and regulate exhaust emissions
	Electricity and industry	Use cleaner fuels Taller chimneys to break through inversion layers Clean up emissions through government regulation
Effects	Smog prevention	Titanium dioxide coated concrete converts NOx to nitrates Design and plan city buildings to promote natural cooling and circulation Promote opening up and cleaning up of covered water courses to allow evaporative cooling
	Health	Raise awareness of conditions and effects of breathing polluted air Promote pollution related health check ups Activated charcoal masks Provide public access to pollution monitoring

Table 6.3: **Management strategies for urban air pollution**

6.3.6 Evaluating Management Strategies for Urban Air Pollution

When evaluating urban pollution, consider the social response to schemes, their economic costs, and their effectiveness.

Here are some points to consider in evaluation:

- Most urban pollution comes from transport, particularly private cars.
- 80% of the car pollution comes from 30% of the cars.
- Diesel engines in trucks and buses produce more particulates.
- Catalytic converters decrease fuel efficiency and increase CO_2 emissions.
- Restrictions and tolls can make car use expensive.
- There may be cultural resistance to public transport.
- Monitoring and regulating is complicated and expensive.
- Groups like WHO set international standards but national standards vary.

Practice questions: Photochemical Smog

1. ***Describe*** the primary pollutant sources and production of tropospheric ozone as a secondary pollutant.

2. ***Outline*** the effects of tropospheric ozone pollution.

3. ***Outline*** how a photochemical smog is formed.

6. ATMOSPHERIC SYSTEMS AND SOCIETIES

4. *Describe* urban air pollution management strategies in the following table:

Level	Strategy	
Cause	Cars, buses, and taxis	
	Electricity	
Release and transfer	Cars, buses, and taxis	
	Electricity and industry	
Effects	Smog prevention	
	Health	

5. *Evaluate* the management strategies you have described.

....

....

....

....

6.4 Acid Deposition

6.4.1 Causes of Acid Deposition

Acid deposition refers to both wet and dry forms of acidic pollution. The most common source is fossil fuel combustion that leads to sulphur dioxide (SO_2) and nitrogen oxides (NO_x) forming. In dry deposition these may form into atmospheric sulphates and nitrates, which can be deposited and cause acidification and eutrophication of ecosystems. Sulphur dioxide and nitrogen oxides combine with any form of atmospheric water to form acid precipitation, most commonly rain containing sulphuric and nitric acids which can have a pH as low as 4: 'acid rain'.

153

6.4.2 Effects of Acid Deposition

	Direct Effects	Toxic Effects	Nutrient Effects
Soil	Some decomposers (fungi and bacteria) and detritivores (worms) are unable to tolerate changes in pH.	Toxic metals such as aluminium, mercury, or cadmium are released from soil minerals by the chemical effects of acid.	Increase in acidity leads to accelerated leaching of calcium and magnesium.
Water	Acids damage exoskeletons of crustaceans such as shrimp and crabs and the jelly of amphibian spawn.	Aluminum ions released into water from soils cause a build-up of mucus on fish gills and death from lack of oxygen.	Nitrate is added to nutrient load in eutrophic waters.
Living organisms	SO_2 and NO_2 in atmosphere damages the tree leaves.	Trees uptake toxic aluminium ions from soils.	Lack of calcium and magnesium leads to reduced ability to carry out photosynthesis.

Table 6.4: **Summary of direct, toxic, and nutrient effects of acid rain deposition**

6.4.3 Regional Impact of Acid Deposition

In contrast with global pollution problems, such as ozone and climate change, or local ones such as urban air pollution, acid deposition problems are regional. Acidified deposition can travel easily across natural boundaries from one nation to another. For example, the Swedish government calculated that only 12% of acid deposition in Sweden was from within the country, the rest came from at least seven neighbouring industrial countries in Europe. Similarly, much of the acid deposition in Canada comes from the USA.

6.4.4 Buffering Capacity of Ecosystems

Different rocks and soils have different capacities to absorb acid deposition, as do the communities found there. Chalk and limestone areas have an ability to buffer acidity as the calcium carbonate neutralises acids. Naturally acidic geology, produced by rocks such as granite, has less capacity to buffer and so the impact won't be reduced. Plant communities also vary in their sensitivity to extra nitrogen. For example, in the UK, low nitrogen communities, such as upland bogs and heath are most at risk.

6.4.5 Management Strategies for Acid Deposition

Level	Strategy	
Cause	Cars, buses, and taxis	The same as for urban air pollution
	Electricity	Alternative energy for power generation; HEP, solar, wind Reduce total demand through awareness raising Reduce demand through better energy conservation
Release and transfer	Cars, buses, and taxis	The same as for urban air pollution
	Electricity and industry	Use cleaner fuels such as low sulphur coal Clean up acidified emissions by using limestone scrubbers (neutralises the acid in the chimney)
Effects	Soils	Liming of soil (adding calcium oxide from baked limestone) to reduce acidity
	Water	Lime bombing of lakes has been effective in increasing pH in Scandinavia Monitoring of plankton indicator species to gauge recovery
	Living organisms	Endangered species of amphibian need captive breeding, e.g. mountain yellow-legged frogs, *Rana muscosa*, in California

Table 6.5: **Summary of management strategies for acid deposition**

6.4.6 Regional Agreements on Trans-boundary Pollution

Agreements can be reached between countries about the causes of acidification and, possibly, compensation. These may be bi-lateral, for example, the 1991 Air Quality Agreement between the US and Canada focused on acid rain and was later expanded to cover smog. Some agreements may cover larger regions, for example, the 1999 Gothenburg agreement to abate acidification, eutrophication and ground-level ozone in Europe.

6.4.7 Evaluating Management Strategies for Acid Deposition

When evaluating strategies for managing acid deposition consider the need for regional agreements between governments, or through regional institutions such as the European Union. Also consider the economic costs and the effectiveness of any approach.

Here are some points to consider in your evaluation:

- Strategies for urban areas may be related; urban pollution includes acid deposition.
- Acid deposition travels with wind and water vapour in the atmosphere.
- Additional environmental impacts of cleaning up emissions, for example, mining, baking, and transporting limestone.
- Lichens can be used as biotic indicators of SO_2 levels.
- Monitoring and identifying sources may be difficult, as they are often non-point.
- Intergovernmental agreements often require proof and appropriate compensation.

ENVIRONMENTAL SYSTEMS AND SOCIETIES SL

Practice questions: Acid Deposition

1. **Outline** the chemistry of acidified precipitation.

   ....

   ....

   ....

2. Complete the following table describing three possible effects, one on water, one on soil, and one on living organisms

Direct effects on:	Toxic effects on:	Nutrient effects on:

3. **Explain** how acid deposition is a regional problem, stratospheric ozone depletion a global problem, and tropospheric ozone a local one.

   ....

   ....

   ....

4. **Describe** pollution management strategies for acid deposition by completing the following table:

Level	Strategy	
Cause	Cars, buses, and taxis	
	Electricity	
Release and transfer	Cars, buses, and taxis	
	Electricity and industry	
Effects	Soils	
	Water	
	Living organisms	

5. **Describe** the importance of international agreements in effectively tackling acid deposition.

　

6. **Evaluate** the strategies you have described.

　

7. Using examples from any form of pollution studied in this unit, **evaluate** the approaches to pollution management at the levels of cause, release and transfer, and effects.

ENVIRONMENTAL SYSTEMS AND SOCIETIES SL

Chapter 7: Climate Change and Energy Production

7.1 Energy Choices and Security

7.1.1 Energy Resources and Global Energy Supply

Global energy consumption can be measured in terawatts (TW). 1 TW can power 10 billion 100-watt bulbs at the same time.

1 TW = 1,000 gigawatts (GW)
= 1 × 10^{12} watts
= 1,000,000,000,000 watts

Estimates have been made that in the year 2000 global energy consumption was round 10 TW, and in 2010 over 16 TW was needed. With a developing and growing population, over 30 TW may be needed by 2100 (Worldwatch Institute, 2009). Global energy supplies are generated from the following sectors (see Table 7.1). Other alternative energies in this case include geothermal, wind, wood, and solar energy.

Year (capacity in TW)	1980 (9.5TW)	2006 (15.8TW)	2016 (17.7TW)
Oil	46	36	33
Gas	19	23	24
Coal	25	27	28
Hydroelectric power (HEP)	6	6	7
Nuclear power	3	6	5
Other alternatives	Less than 1	1	3

Table 7.1: **Global energy supply (%) in 1980, 2006 and 2016**
Source: Adapted from EIA, 2008

Fossil fuels continue to supply the vast majority of total primary energy supply in all sectors, with the gas and coal sectors expanding in their contributions. Note that there has also been a significant increase in the contribution of alternatives in the 21st century. Total energy requirement continues to increase, with demand being met by growth in coal, gas, and alternatives. Nuclear and oil remain constant, so the percentage declines.

You need to be able to outline the various energy sources available to society as shown in Table 7.3 on page 160.

7. CLIMATE CHANGE AND ENERGY PRODUCTION

7.1.2 Electricity Generation and Carbon Dioxide Emissions

Fossil fuels release carbon dioxide when combusted; when used for energy supply they have higher emissions than other sources. Emissions below are given per kWh of electrical energy produced; in other situations, the order may change.

All energy sources have some greenhouse emissions, though, for systems such as solar (photovoltaic) the emissions are indirect from other stages in the life cycle not from use. Biofuels are a special case, as they are based on living plants and have to absorb the carbon dioxide from the atmosphere before they are burnt. However, there may be some carbon emissions from production of biomass, for example, fertiliser and machinery used. Nuclear power can have very low emissions from the production plant's complete lifecycle; however, it remains a controversial choice due to problems associated with radioactive waste and the risk of accidents.

Energy Source	CO_2 per kWh
Coal	1,000
Oil	840
Gas	470
Solar	48
HEP	24
Wind	12
Nuclear	12

Table 7.2: **Carbon dioxide released per kilowatt hour for different energy sources**
Source: includes data from BP, 2017

7.1.3 Factors That Affect Energy Sources Adopted by Different Societies

You should be able to discuss the choice of different energy sources for any society. Table 7.3 summarises some of the factors that influence that decision, but this needs to be brought into the context of the country and its society. Consider the following general statements that influence the decision made:

▶ **Availability**: resources within or near to a country are better than those further away.

▶ **Economic**: costs include capital costs and running costs. High capital with low running costs may not be affordable to poorer countries or may require loans to be taken out.

▶ **Cultural**: energy forms that have been used previously are often preferred within that culture (this resistance to change is called 'cultural inertia').

▶ **Environmental**: legislation, education, and dominant environmental value systems may influence countries to consider the environmental impacts in different ways.

▶ **Technological**: higher technology may require expensive training of the workforce, or import of foreign workers to advise.

▶ **Security**: energy that promotes independence is favoured, where it is affordable and reliable for the society. Dependence on energy supplies purchased from uncertain allies represents a security risk, for example, Russia cutting the supply of natural gas to Ukraine.

▶ **Sustainability**: increasingly a priority for energy security, but one only ultimately favoured by renewable sources, as all fuel based production systems will eventually deplete the fuel source.

Consider these factors in relation the table of advantages and disadvantages, and in the case study given.

Fuel Type	Advantages	Disadvantages
Oil	Extremely versatile Relatively easy to transport High energy content against mass	Rising costs Air pollution: nitrogen oxides, carbon dioxide Location of resources in only a few countries Peak production near
Natural Gas	Versatile Cheaper than oil Fair energy content against mass Low emissions on combustion	Inflammable Non-renewable Impact of mining and pipelines
Coal	Will last longer than oil Cheap in comparison with oil Electric power generation can be built near to large coal fields to give cheap power	Air pollution from combustion particulates, sulphur dioxide, and carbon dioxide. Mining can be difficult, dangerous, and polluting Non-renewable Bulky and hard to transport
Hydroelectric power	Low quantities of air pollution (some from decomposing plants) Renewable Provides reservoir water for fishing, recreation, and irrigation	Depends on adequate rainfall and river flow May displace riverside communities Destroys existing natural ecosystems by submerging them
Nuclear fission power	Does not produce carbon dioxide Requires small quantities of fuel Can produce large quantities of electricity	There are limited supplies of suitable fuel, e.g. Uranium Radioactive wastes may pollute the environment, for a long time There is a risk of 'meltdown'
Solar	Continuous supply Converted to electricity by photovoltaic cells Can be used directly for cooking, heating water or houses	Large seasonal variation at high latitudes Large areas of land are needed to produce electricity On average the most expensive option
Geo-thermal	Continuous supply possible Relatively low air pollution—no fuel burnt Costs low in the long term	Requires specialist training and equipment Short term initial capital costs may be expensive Only suitable in certain areas (near plate boundaries)
Wind farms	Renewable Low pollution mainly due to manufacture Similar cost to thermal power stations	Visual and noise pollution in beautiful areas Hazard to wildlife Only suitable for windy areas
Biofuels	Often traditional fuel CO_2 neutral (theoretically) Can provide habitat for wildlife	Destruction of natural habitats Irrigation demands Can compete with food supply

Table 7.3: **General advantages and disadvantages of energy sources available to society**

7.1.4 Energy Efficiency and Conservation

One option to deal with increased demand for power is to meet or reduce the demand through an increased efficiency of use. This may be achieved by promoting energy conservation, improved grid design, and buying back surplus from buildings that generate their own electricity. There can be considerable cultural inertia to such demand side management.

7.1.5 Case Study of Electrical Power Generation for Thailand

Thailand's electricity is supplied by the Electricity Generating Authority of Thailand (EGAT) to the Metropolitan Electricity Authority (MEA) who supply the power to the consumer's house. Table 7.4 shows the methods that EGAT used in 1998 to meet electricity supply needs, and Figure 7.1 shows the predictions that they made for future demands.

Type of Generation	Fuels Used	MW	%
Hydropower		2,874	16
Thermal power	Coal (some Oil and Gas)	6,517	36
Combined cycle power	Gas	5,074	28
Peaking plants	Gas	892	5
Power purchase	Import from Laos (HEP) or Malaysia (Gas)	2,818	15
	Total	18,175	

Table 7.4: **Electricity Generation in Thailand**
Source: Adapted from Thailand Development Research Institute, 2000

> **Local to global**
> Case study examples in this guide allow you to test and apply theory within a real-life context. Ideally you should support your revision with case study examples from your local area or country. Local studies from other countries also help to give perspective on international environmental agreements on global issues.

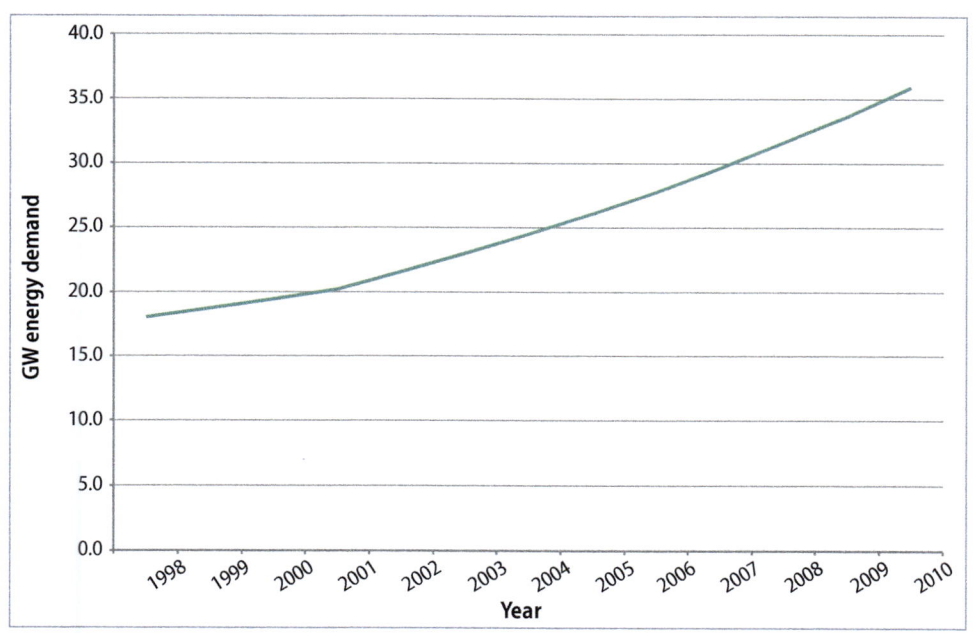

Figure 7.1: **Graph of electricity consumption**
Source: Data from Thailand Development Research Institute, 2000

An energy conservation scheme introduced in Thailand in 1992 saved over 700 MW in eight years at a cost of 189 million US dollars (Fahn, 2004). Compare this 'negawatt' scheme with the two-megawatt production schemes shown in Table 7.5.

	Hydroelectric power (HEP)	Coal power station	Nuclear power station
Figures based on	The Pak Mun Dam, North East Thailand (built 1994)	A proposal to build a coal fired power station at Bor Nok, southern Thailand	A proposal to build two nuclear power stations in Thailand for 2027
Electricity produced	136 MW (although 21 MW actually produced due to low flow and the need to raise sluices for fish)	1,468 MW	2 x 1,000 MW
Fisheries	Increased production from the reservoir fisheries (estimated at 320,000–693,000 US $)	May have negative influence on squid fishery, not a positive gain	Possible radioactive contamination from waste and coolant water
Cost (US $)	260 million	820 million	1.5 billion
US $ per MWh	77	84	79
Social costs (including government paid compensation)	1,700 households evacuated (compensation = 36 million US $) 6,202 households affected by damage to the river fisheries (compensation = 15 million US $)	The whole community is influenced by damage to agriculture (pineapple growing) and the squid fisheries Thermal pollution as seawater is used as a coolant. Coal dust likely from the ships unloading at the pier	Nuclear power in consideration since 1966 in Thailand Public opposition is strong, with demonstrations in proposed sites in Kalasin and Surathani
Water pollution	The acidity of water increased due to decomposition, both in the lake and downstream from the dam	Sulphur dioxides (would be cleaned by limestone scrubbers)	Risk of radioactive contamination with potential to damage tourism and fisheries
Air pollution	Some greenhouse gases produced from decomposing vegetation at the bottom of the reservoir	Particulate air pollution Massive amounts of greenhouse gases, especially carbon dioxide	
Soil	Productive land damaged by flooding in the area above dam	Soil likely to be polluted by the air pollution	
Ecological effects	Damage to the migration of fish and any natural vegetation in the flooded area Species extinctions estimated at 50 species, with a decline in 169 species of fish Damage to the fishery estimated at 1.5 million US $ per annum, a 60–80% decline	Thermal pollution of sea due to the water cooling systems Air pollution to surrounding forest is likely Whales, dolphins and rare water birds to be affected Fish migration influenced by large pier	Very low carbon dioxide emissions, represents a 15% cut
Additional costs	Forest losses includes 40 edible species of plants, 10 species of bamboo, and 45 mushroom species Medicinal plants also lost	Coal has to be imported from Indonesia Coal mining and limestone quarries have additional environmental impact	Nuclear fuel needs to be sourced

Table 7.5: **A case study based comparison of three energy sources in Thailand**

Source: Thailand Development Research Institute, 2000; Matsumoto & Fukuda, 2002; IEA, 2016

7. CLIMATE CHANGE AND ENERGY PRODUCTION

Practice questions: Climate Change and Energy Production

1. ***Outline*** the range of energy sources available to society.

   ....

   ....

   ....

   ....

> **Renewable doesn't mean recycled!**
>
> Be very careful if you use the words recycling and energy in the same sentence or paragraph. It is easy to contradict the laws of thermodynamics.

Look at the table below comparing global energy supply with Thailand (population 68.14m) and the UK (population 61.14m).

Year (Production Capacity)	Global (15.8TW)	Thailand (35GW)	UK (60GW)
Oil	36	1	1
Gas	23	33	41
Coal	27	35	33
Alternatives	7	16	3
Nuclear power	6	0	19
Imported		15	3

2. ***Compare and contrast*** the energy choices made by Thailand and the UK.

   ....

   ....

   ....

   ....

3. ***Suggest*** reasons for the differences in capacity of nuclear power and alternatives between Thailand and the UK.

   ....

   ....

   ....

   ....

   ....

4. *Suggest* reasons for the difference between energy production capacity from oil globally and within these two countries.

....

....

....

....

5. *Compare and contrast* the advantages and disadvantages of HEP, coal, and nuclear using specific examples.

....

....

....

....

6. *Outline*, for a stated example of one country, the factors that influence the choice of energy source.

....

....

....

....

7.2 Climate Change: Causes and Impacts

7.2.1 Climate and Weather

Weather is the day-to-day changes in the atmosphere around us, including precipitation, temperature, humidity, wind speed, and cloud cover. Climate is the average weather for an area over a full range of seasons. The most important measures of climate type are temperature and precipitation, influenced by atmospheric and oceanic circulation as determined by geographical location. Climate change means a long-term shift in average conditions for a given area, resulting in changes in weather experienced on a day-to-day basis. This has occurred throughout the history of the Earth, but current climate change has been clearly attributed to human activity by climate scientists.

7.2.2 The Role of the Greenhouse Effect in Maintaining Mean Global Temperatures

See section 6.1.

Gases such as carbon dioxide, water vapour, and methane occur naturally in the Earth's atmosphere. These gases are called greenhouse gases (GHGs), as they act in a similar way to the glass panels of a greenhouse, only on a global scale. The GHGs allow short wavelengths of radiation, such as visible light and UV, to pass through to the Earth's surface but they absorb the longer wavelengths, such as infrared radiation, that are radiated from the earth. This is called the 'greenhouse effect', and it regulates the Earth's temperature keeping conditions warmer than without the effect and more hospitable to life.

Carbon dioxide levels and temperatures have fluctuated over time. The atmospheric conditions in the earliest record (3500 mya) indicate changes occurring when an oxygen-rich atmosphere was first created. It should be remembered that comparisons with records from longer ago represent a world with very different solar inputs, climatic and oceanic circulation and living organisms.

> See plate tectonics and biodiversity in section 3.3.2.

The most useful comparisons with today's climate from the geological past are the temperature and carbon dioxide fluctuations of the last two million years. This period includes several long ice ages and warmer shorter interglacial periods. The maximum and minimum from these periods is shown in Table 7.6, details can be found in graphs of ice core data.

Time period	3500 mya	450 mya	350 mya	200 mya	Glacial min	Inter-glacial max	Pre-industrial	1959	1990	2000	2011	2017
Method	From geological determination				From ice core data			Values of CO_2 measured Mauna Loa, Hawaii				
Temperature change* (°C)	+10	+2	0	+4	-8.5	+3.6	0	+0.2	+0.3	+0.4	+0.6	+1.1
Carbon dioxide (ppm)	7,000	4,200	800	1,700	180	300	280	316	354	370	392	410

Table 7.6: **A summary of past and recent variation in CO_2 and temperature °C (mya = million years ago)**
*temperature change from pre-industrial levels
Source: Includes data from Monbiot, 2007; Stern, 2006; IPCC, 2007; and IPCC, 2013

7.2.3 Human Activities and Greenhouse Gases

Table 7.7 summarises the main anthropogenic greenhouse gases in the atmosphere. Gases are rated in terms of their warming potential: the per molecule radiative forcing (heat trapping) capability per molecule of gas. Carbon dioxide is used as the benchmark of this with potential of 1. Although this is very low per molecule, there is such a high concentration of CO_2 compared with the other greenhouse gases that it remains the most significant greenhouse gas whose concentration has been changed by human activity. It is an accepted fact that the global average concentrations of the main greenhouse gases shown in the table below have increased due to human activities over a short time scale.

>
> CO_2e
> CO_2 is used as a baseline measure and the unit of international regulation is carbon dioxide equivalent—CO_2e, based on the warming potentials described in Table 7.7.

Gases	Source	Significance
Carbon dioxide, CO_2	Combustion, decomposition, or respiration Caused by damage to ecosystems, e.g., deforestation and combustion of fossil fuels	No. 1 cause of ACC* Potential = 1 Atmospheric concentration = 410 ppm**
Methane, CH_4	Anaerobic decomposition in rice fields and landfill Cows and stores under the arctic permafrost and oceans	No. 2 cause of ACC Potential = 34 *** Atmospheric concentration = 1.8 ppm
Nitrogen oxides, NO_x	Internal combustion engines	No. 3 cause of ACC Potential = 289 Atmospheric concentration = 0.33
Chlorofluorocarbons (CFCs, also HCFCs and HFCs)	From refrigeration coolants and aerosol cans	Most potent GHG per molecule Potential for HFC-23 = 12,000

Table 7.7: **Summary table of the main greenhouse gases that are altered by human activities**

* ACC = anthropogenic climate change

** ppm = parts per million

*** global warming potentials are CO_2e for a 100 year period per molecule of the greenhouse gas. Methane has a far greater short term effect on release of 150-200 times that of carbon dioxide

Source: IPCC, 1990; IPCC, 2007; IPCC, 2013; Henson, 2011; Page, 2011; Beckwith, 2019; Etminan et al., 2016

7.2.4 Focus on Carbon Dioxide

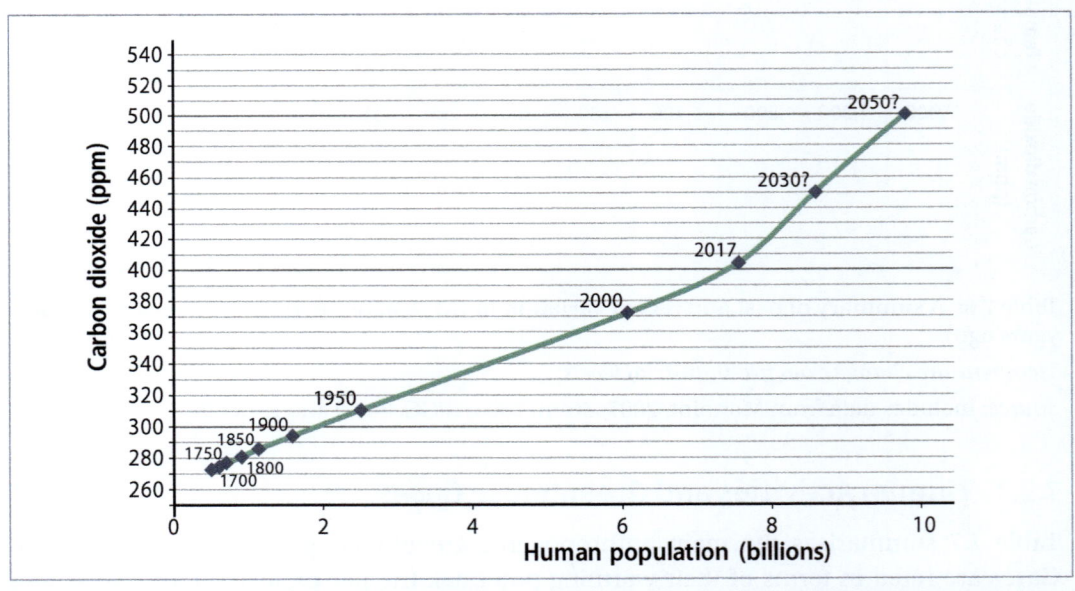

Figure 7.2: **Summary table of the main greenhouse gases that are altered by human activities**
Source: United Nations, 2008; IPCC, 2007 and 2013

The trend shown by carbon dioxide emissions is related to all human activities involved in managing natural ecosystems, agriculture, and energy use. The graph shows changes in relation to 50-year periods in population and in carbon dioxide. The greatest amount of change in CO_2 levels on the graph occurred from 1950–2000 but greater change is predicted in the period from 2000–2050.

7. CLIMATE CHANGE AND ENERGY PRODUCTION

7.2.5 The Special Case of Water as a Greenhouse Gas

The role of water in climate regulation is complex. As water vapour, it is the most significant greenhouse gas in the natural greenhouse effect and is responsible for over 36% of natural warming. However, water is also capable of a cooling effect due to increasing reflection (albedo) from cloud surfaces. Water condensation and evaporation through the hydrological cycle are important for both heating and cooling on a local and a global scale. Currently, there is comparatively low scientific understanding about the role of water vapour, both as a greenhouse gas and as a potential solution. For these reasons, scientists consider water to be a feedback agent, rather than a forcing factor of climate change.

7.2.6 Enhanced Greenhouse Effect and Anthropogenic Climate Change (ACC)

Climate change is a natural occurrence and both gradual and sudden variations in temperature punctuate the fossil record. The term 'climate change' can be used to describe global warming. Anthropogenic (human generated) climate change is the widely accepted theory that the current global warming is caused by an enhanced greenhouse effect, which has been created by human emissions altering the concentrations of both naturally occurring and manufactured greenhouse gases.

Global warming is a term used to describe an average increase in the planet's temperature. Such a warming trend has been observed and measured in both historical records and from studies using various methods (for example, ice cores, tree rings, and indicator species of plant pollen or beetle remains) which can go back into prehistory. The temperature changes that have been observed are generally accepted as valid indications that the Earth warmed during the 20th Century and is continuing to do so at the start of the 21st.

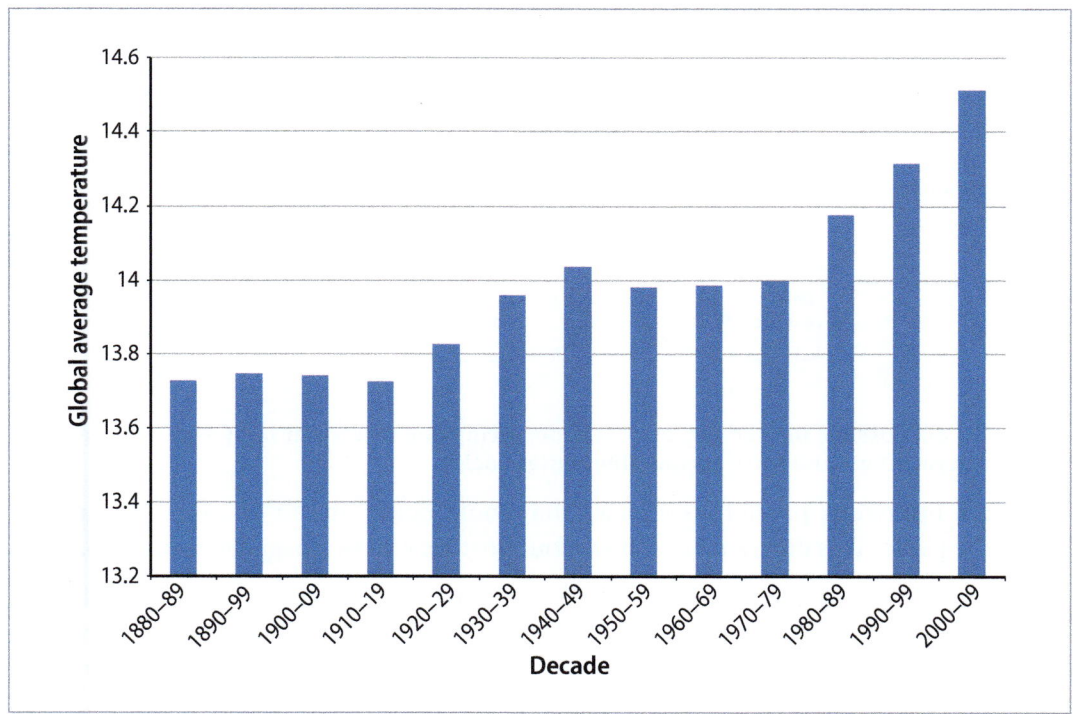

Figure 7.3: **Decade averages of global average temperatures from 1880–2010 from historical records**
Source: Data from NASA

Predicting future climate change is done using computer based global climate models (GCMs) by climate scientists. There is considerable uncertainty over the predictions made by these models due to the complexity of the systems, the prediction of human behaviour needed, and the number of feedback loops involved. See Figure 7.4 for some examples.

7.2.7 Some Feedback Mechanisms Associated with an Increase in Global Temperature

Tipping points and stability were discussed in section 1.3.10 on page 16. It may be useful to look back to the discussion there and review Daisyworld in 1.3.11.

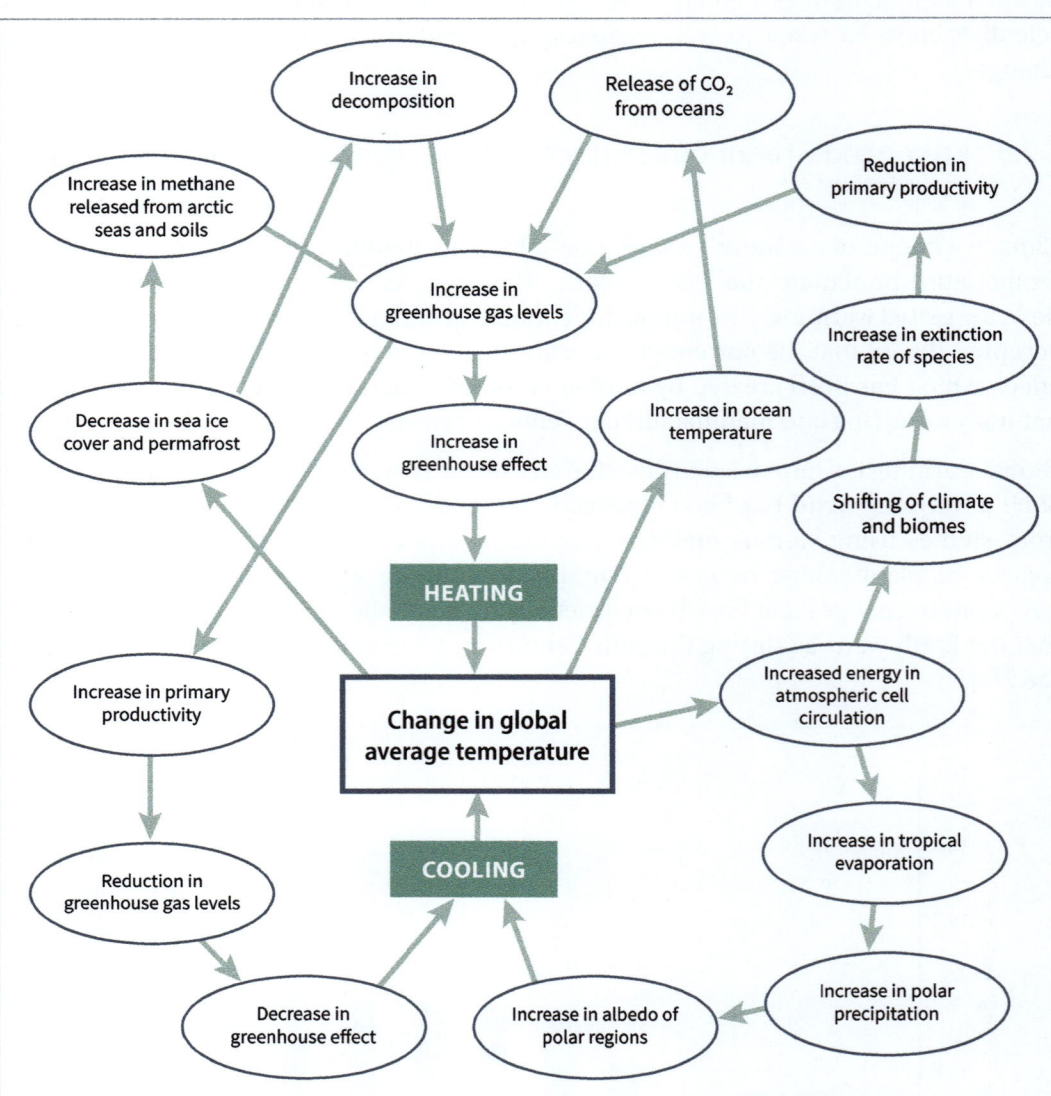

Figure 7.4: **Six possible interacting feedbacks occurring due to warmer temperatures (start at the central rectangle and trace six possible routes back.)**

There are numerous possible feedbacks that could occur due to an average warming across the globe. This diagram only shows four positive and two negative feedbacks.

Although the arguments on ACC are often presented as simple cause and effect, feedbacks show that the factors change together. In the case of CO_2 and temperature, either can lead the change and both have been seen to do so in the past. Feedbacks take place on long timescales and create considerable uncertainty in the GCMs.

One issue is that of tipping points or run-away effects. If certain thresholds are reached then positive feedbacks may be impossible to regulate or reverse, for example, the release of methane from melting of Arctic permafrost and sea ice.

7. CLIMATE CHANGE AND ENERGY PRODUCTION

7.2.8 The Intergovernmental Panel on Climate Change (IPCC)

The IPCC brings together scientists, politicians and other interested parties in three working groups:

- Group 1: the scientific aspects of the climate system and climate change
- Group 2: the consequences and response of society to climate change
- Group 3: options to reduce greenhouse gas emissions

The predictions made for climate change are based on scenarios of human behaviour over the 21st century. Predictions range from a warming of 2°C to 6°C by 2100. Details are published online at http://www.ipcc.ch. These predictions are based on the fourth and fifth assessment reports (IPCC, 2007 and 2013). Since the time of writing, AR6 has been published (IPCC, 2021) and for the most up-to-date information consulting this is recommended.

7.2.9 Consequences of Climate Change

These changes may not seem like very large alterations, but in the last interglacial temperatures were 3.6°C higher and sea levels over 4 m higher than today. The consequences of the predicted change vary considerably from one location to another. Changes in local weather may even benefit some areas as colder countries can grow more food with a mild warming. However, most of the predictions are not for beneficial change but for an increase in adverse conditions for society as a whole—particularly if the higher levels of predicted warming are reached.

Changes include:

- Warmer conditions and increase in heatwaves in most land areas
- Heavier precipitation events in most areas
- Increase in droughts
- Increase in the frequency and intensity of cyclones (hurricanes and typhoons)
- Biomes shifting latitude and altitude
- Crop growing regions shifting
- Coastal areas flood due to thermal expansion of oceans and land-based ice caps melting
- Tropical diseases spread to higher latitudes.

7.2.10 Contrasting Human Perceptions

You should contrast the views of others with your own on the main debates.

Scientific debate continues on ACC. The most reliable source for the consensus viewpoint on the science remains the IPCC as summarised here. Evidence of consensus is clear amongst specialist researchers according to surveys of the peer reviewed literature; see 2010 and 2004 in the timeline.

However, it should be remembered that a consensus does not necessarily equal the truth. There are a range of arguments put forward against ACC suggesting essentially that the climate has always changed and that the current change may not be anthropogenic. These arguments often focus on the inherent uncertainty of the models, the role of natural factors, and the difficulty of predicting the future using scenarios (different models of societies in the future).

Degrees Celsius of Warming (Possible Dates)	Possible Consequences
1 (2015–2040)	• Increase in desertification • Expansion of tropical belt areas and associated diseases • Increase in bleaching events in coral reefs
2 (2050–2100) Working Target for UN*	• Severe shortages of water widespread • Amazon rainforest significantly damaged • Annual heat waves experienced in temperate zones • Greenland ice sheet could melt
3 (2080+)	• Open ocean at the North Pole • Global food production significantly damaged • Category 6 cyclones in tropics • Rainforest become savanna and desert
4 (2100+)	• West Antarctic ice sheet collapses • Desert conditions develop in Southern Europe • Permafrost in Northern Hemisphere gone
5 (2100+)	• Temperatures in the Arctic Ocean typical of today's Mediterranean • Large scale migrations of human population likely • Subtropics uninhabitable
6 (2100+)	• Uninhabitable areas expand globally • Potential collapse of human civilization • Mass extinction

Table 7.8: **Potential consequences of predicted warming**

At Copenhagen in 2009 (COP15) it was agreed that warming should not go above this level, later lowered to 1.5°C

Source: Monbiot, 2007; Lynas, 2008; Stern, 2006; and IPCC, 2007

The perception of the general public in the US and the UK is that the science is divided and trust in scientists has declined (Tollefson, 2010). Surveys have shown how public opinion has altered. UK citizens were asked, 'is the world's climate changing?'

2010: 78% of 1.822 people said yes; 15% said no.

2005: 91% of 1,491 people said yes; 4% said no.

(Spence, Venables, Pidgeon, Poortinga, & Demski, 2010)

US citizens were asked, 'do you think global warming is happening?'

2010: 57% of 1,001 people said yes; 20% said no.

2008: 71% of 2,189 people said yes; 10% said no.

(Leiserowitz, Maibach, & Roser-Renouf, 2010)

In the UK 2010 survey, 78% of those questioned thought the climate change was mostly due to human activities or a mixture of human and natural causes, with 18% saying it is mostly due to natural causes. In the US 2010 survey, 53% believed it was at least in part human caused, with 36% saying it was due mostly to natural causes.

7. CLIMATE CHANGE AND ENERGY PRODUCTION

The nature of climate change skepticism is complex; it is useful to distinguish the following when describing positions on climate change:

1. Trend skepticism (it isn't happening).
2. Attribution skepticism (we are not causing it).
3. Consequence skepticism (it's not going to be a problem).
4. Management skepticism (we can't do anything about it anyway).

The most significant factor influencing our position on climate change is our existing worldview. It is clear that our EVSs have a large influence on our beliefs. Major influences on the world view and therefore outlook on climate change include religious and cultural background, political leanings, age, and gender (Hamilton, 2011; Monbiot, 2007; Klein, 2014).

For further information see Rahmstorf (2004).

7.3 Climate Change: Mitigation and Adaptation

7.3.1 Mitigation and Adaptation

These are two very different and contrasting responses to climate change. Mitigation is the attempt to reduce the cause of the problem through reducing GHGs or at least stabilising them. A look at the timeline of events in relation to climate change shows that this has been the usual approach. Mitigation strategies include consideration of energy supplies and reducing total energy consumption, reducing emissions of other GHGs in other sectors, for example, methane from cattle and rice farming. Carbon dioxide removal techniques are also proposed, to protect and enhance natural carbon sinks such as forests and oceans.

Adaptation strategies are focused on living with the effects of climate change, through reducing the impacts on societies. Essentially these are looking at coping with the effects of climate change on a local level. These include most strategies towards greater sustainability discussed in this course.

Further details of these management strategies are found in Table 7.9. You should learn at least two adaptations and two mitigation strategies in depth.

A third and overlapping category exists: geoengineering. This is an attempt to manipulate the climate on a global scale. For example, to stop total loss of arctic ice, a 'blue ocean event', some climate change researchers suggest that immediate geoengineering such as solar radiation management (Ricke, et al., 2012) or cloud brightening (AMEG, 2014) is required now.

7.3.2 International Efforts to Deal with Climate Change

The United Nations Framework Convention on Climate Change (UNFCCC) is an international treaty adopted during the Earth Summit at Rio de Janeiro in 1992. This group debates political agreements on climate change as advised by the IPCC which itself was appointed by the group. The progress of this group is detailed in the timeline on page 173. Much of the political debate is now focused on policies of how to achieve agreed targets through mitigation. UNFCCC has also set up National Adaptation Programmes of Action (NAPAs) for the LEDCs that may require assistance from MEDCs to adapt to climate change.

Debate on management strategies is widespread, both as part of national policy, intergovernmental agreements, or personal lifestyle choices. In UK surveys, 71% agreed that they had a responsibility to act and 63% agreed that they could personally help by lifestyle change. National government (32%) or the international community (30%) should be mainly responsible for taking action.

To many, the precautionary principle is seen as the wisest course of action. There will never be evidence to prove without a doubt that human emissions cause climate change (it just seems to be very likely to a lot of climate scientists). However, the lack of proof doesn't matter as there is no evidence that human emissions don't cause global warming. As we have to then accept that there is a very good chance that the emissions might be the cause of current and future warming, and that it could be disastrous for both people and ecosystems, then we should do something anyway. To continue to increase emissions can only be seen as the logic of gambling with the future.

7.3.3 Some Management Strategies for Climate Change

Strategies for dealing with climate change fit into three categories: mitigation, geoengineering, and adaptation.

Level	Strategy	
Mitigation (preventive measures to reduce anthropogenic emissions of known GHGs)	Carbon taxes	Require emitters to pay a fee for every tonne of GHG emitted. This is already implemented in several countries, e.g., Sweden and India.
	Carbon trading	Countries or companies emitting above the target level can buy carbon storage credits from clean developments or reforesting degraded land in other countries. Schemes exist under Kyoto Protocol and Copenhagen Accord.
	Cap and trade	Permits to pollute above certain levels are sold on the free market; any organisation that is under allocation can make profit by selling the extra permits.
	Lifestyle changes	Encourage individual actions to reduce climate change, including choices of transport, energy use, and consumer goods and services.
	Carbon dioxide removal	REDD agreement (reduce emissions from deforestation and forest degradation) to encourage tree planting. Ocean seeding, by addition of limiting nutrient such as iron, to promote plankton productivity.
Geoengineering (manipulating environmental systems on a global scale to reduce incoming solar radiation or the greenhouse effect)	Solar radiation management	For example, releasing atmospheric sulphates on a scale equivalent to large volcanic eruption or cloud seeding using sea water.
	Carbon dioxide reduction	Develop technologies to extract GHG from the atmosphere and store them.
	Carbon capture and storage	Carbon emissions from power plants or industry are directed by back into stable geological structures where they are trapped.
Adaptation (changes society can make to deal with the adverse effects of climate change)	Building design	Improve air conditioning and circulation in buildings in the temperate zone.
	Emerging diseases	Monitor and control the spread of spreading tropical diseases.
	Coastal management	Improve sea defences or manage retreat from low lying coastal areas.

Table 7.9: **A few of the many strategies discussed to manage climate change**

7. CLIMATE CHANGE AND ENERGY PRODUCTION

CLIMATE CHANGE EVENTS AND AGREEMENTS

1988	Scientist Dr. James Hansen states that NASA is 99% certain that warming is not natural variation but caused by a build-up of carbon dioxide. The IPCC is established.
1990	The first IPCC report concludes warming has happened and is likely to continue.
1991	Mount Pinatubo erupts, releasing aerosols that cool the climate. Sunspot cycle hypothesis is proposed by Lassen.
1992	UN Framework Convention on Climate Change founded at Rio de Janeiro in 1992; this develops into the Kyoto Protocol.
1995	The second IPCC report concludes that warming has a human 'signature' and declares serious warming to be likely.
1997	Kyoto Protocol sets binding targets for 37 industrialised countries and the European community to cut CO_2e emissions by 5% from 1990 levels. Developing countries had differentiated non-binding targets set.
2000	Evidence for sunspot cycle weakens; Lassen concludes they cannot explain warming.
2001	The third IPCC report states a global warming that is unprecedented since the end of the last ice age and is likely to have serious consequences. Global dimming connected to atmospheric pollution, pollution is masking the warming.
2003	Deadly heat waves occur in Europe.
2004	Research analysing 928 abstracts from scientific journals finds no disagreement with the consensus on ACC (Oreskes, 2004).
2005	Kyoto treaty in effect for major industrial nations except US. Hurricane Katrina spurs debate in America.
2006	Al Gore's *An Inconvenient Truth* wins public opinion globally. 'Hockey stick' climate graph creates controversy and polarises opinion.
2007	The fourth IPCC report finds observed warming is 'very likely' due to anthropogenic GHG concentrations. Bali action plan identifies support for adaptation and reducing emission from deforestation and degraded land (REDD) as areas for action.
2009	Warming occurring faster than predicted. Public confidence in scientific consensus and belief in ACC falls. Copenhagen accord recognises that warming should be kept under 2°C but fails to reach expectations to provide an effective extension of Kyoto.
2010	Cancun climate discussions revive optimism in the process but reach no conclusions on a new protocol. The work of 1,372 climate researchers and their publications screened; 97-98% of researchers support ACC (Anderegg, Harold, Scneider, & Prall, 2010).
2014	The fifth IPCC report concludes the evidence is 'unequivocal' and that even if the world begins to moderate greenhouse gas emissions, warming is likely to cross the critical threshold of 2°C by the end of this century.
2015	Paris Agreement: it is agreed that each country is to determine and plan for its own contribution to mitigate climate change. A non-binding agreement to aim to keep warming under 2°C, with an encouragement to aim for 1.5°C.
2018	Special Report on Global Warming of 1.5°C published by the IPCC and becomes the working target for safe climate change. The world is predicted to reach 1.5 degrees of warming between 2030 and 2052 (IPCC, 2018), though advocates of abrupt climate change suggest it might happen even sooner. Based on the Paris Agreement's unconditional pledges, the world is heading for a 3.2°C temperature rise (UNEP, 2021).

2020	Under the COVID-19 pandemic restrictions greenhouse gas emissions fall by 6.4%, mostly due to a decline in aviation. This level of reduction would have to be exceeded cumulatively each year for the next decade to meet the Paris Agreement (Tollefson, 2021).
2021	The sixth IPCC assessment report on the physical science concludes that anthropogenic climate change is widespread, rapid and intensifying. 1.5 degrees can only be reached through immediate, rapid and large-scale reductions in GHG emissions (IPCC, 2021). The UN Secretary General introduces this report as "a code red for humanity. The alarm bells are deafening, and the evidence is irrefutable: greenhouse gas emissions from fossil fuel burning and deforestation are choking our planet and putting billions of people at immediate risk." (United Nations, 2021).
2021	Analysis of over 88,000 published research papers since 2012 finds that 99% of them support the theory of anthropogenic climate change (Lynas et al, 2021).
2021	COP 26 in Glasgow attracts widespread media attention on climate change and leads to the Glasgow Climate Pact, which includes the first explicit reference to reducing fossil fuel subsidies. Coal is to be "phased down", rather than "phased out" following final edits to the agreement by India and China. Leaders from 120 nations agree to halt and reverse deforestation by 2030.
2021	Climate change causes extreme weather-related events to impact countries across the world (Cockburn, 2021).

The house features passive cooling due to convection currents, solar chimney (top right), geothermal and pond-cooled air intake (bottom and right). The house is painted white for high albedo and the walls are insulated. Note: shady trees attract wildlife.

Figure 7.5: **The author's Bangkok Eco guesthouse**

7.3.4 Evaluating Climate Change Strategies

Climate change management strategies are complex; you will need to consider the role of international agreements and the relative importance of lifestyle change.

7. CLIMATE CHANGE AND ENERGY PRODUCTION

Consider some of the following points:

- A fully ratified Kyoto Protocol would achieve a reduction of warming by around 0.5°C.
- Concerns about fair target setting between MEDCs and LEDCs continue to cause disagreement.
- For a reasonable chance to keep under 2 degrees of warming, levels of around 400–450ppm should be maintained.
- Emission cuts of over 80% CO_2e are needed to achieve 400–450ppm.
- International agreements affect large numbers of people.
- Countries may not sign or agree to international agreements.
- Concerns about the economic cost and impacts on development are widespread.
- The Kyoto mechanism can be seen as unfair as it sets targets against a historical baseline (1990), so countries that had high emissions then can maintain them.
- Estimating carbon storage in forests involves predictions about the future of those forests.
- Carbon storage in ecosystems in not well understood and difficult to monitor.
- Few individuals are prepared to make significant cuts in lifestyle related emissions independently.
- Simulated volcanic eruptions are unpredictable and could damage the ozone layer.
- Cloud seeding could inadvertently cause an increase in the greenhouse effect.
- Climate scientists are increasingly exploring geoengineering options as emission mitigation attempts over two decades appear to have failed.
- Adaptation methods do not require international co-operation.

Practice questions: Mitigation and Adaptation

1. **Describe** the role of greenhouse gases in maintaining global temperature.

....

....

....

....

2. **Describe** how carbon dioxide has varied in geological time.

....

....

....

....

3. Complete the following table on greenhouse gases.

Gases	Source	Significance
		No. 1 cause of ACC Potential = 1 Atmospheric concentration = 410 ppm
		No. 2 cause of ACC Potential = 34 Atmospheric concentration = 1.8 ppm
		No. 3 cause of ACC Potential = 289 Atmospheric concentration = 0.33
		Most potent GHG per molecule Potential for HFC-23 = 12,000

4. ***Describe*** how human activities add to greenhouse gases.

....

....

....

5. ***Describe*** the evidence for climate change over the last 100 years.

....

....

....

6. ***List*** three methods for investigating climate change before human records began.

....

....

7. ***Construct*** two flow diagrams: one showing a positive and one a negative feedback involved in a global warming.

8. With reference to the quote:

   > *Many critiques somehow understress the fact that the sword of uncertainty has two blades: that is, uncertainties in physical or biological processes which make it possible for the present generation of models to have overestimated future warming effects, are just as likely to have caused the models to have underestimated change.*
   >
   > Stephen H. Schneider (Schneider, 1990)

   Discuss the difficulties associated with predicting climate change, including the role of feedbacks.

   ....
   ....
   ....
   ....
   ....
   ....
   ....

9. Complete the table below to show the potential consequences of predicted warming during the 21st Century:

Degrees Celsius Of Warming (Possible Dates)	Possible Consequences
1	
2	
3	
4	
5	
6	

ENVIRONMENTAL SYSTEMS AND SOCIETIES SL

10. Consider the targets set at Rio, which developed into the Kyoto Protocol: a reduction of approximately 5% of 1990 levels of greenhouse gas emissions. What was actually achieved: an increase in carbon dioxide emissions of around 9% from 1992–2001. Later this fails to develop into a binding agreement in Copenhagen or Paris.

    Explain why carbon dioxide emissions are so hard to control.

    ....

    ....

    ....

    ....

11. *Compare and contrast* the success of the Montreal and Kyoto protocols.

    ....

    ....

    ....

    ....

12. *Describe* pollution management strategies for ACC against the following headings:

Level	Strategy
Mitigation	
Geoengineering	
Adaptation	

13. *Evaluate* pollution management strategies at the three levels in the table.

    ....

    ....

    ....

    ....

    ....

7. CLIMATE CHANGE AND ENERGY PRODUCTION

14.
> *I would like to believe that the changes I suggest could be achieved by appealing to people to restrain themselves. But…self-enforced abstinence alone is a waste of time.*
>
> George Monbiot (Monbiot, 2007)

Evaluate the significance of lifestyle change and international agreement in influencing climate change.

....

....

....

....

....

....

15.
> *There is an anti-science group, a flat earth group over the evidence that exists about climate change.*
>
> Gordon Brown 2009, UK Prime Minister

> *Science does not function by consensus, most certainly not by politically driven consensus*
>
> Prof. Philip Stott 2009 (in response to above),
> Professor of Biogeography, University of London (AmyBLUF, 2009)

Compare these quotes and viewpoints presented in the questionnaire data from the US and the UK, with your own viewpoint. *Explain* why people have such varied perceptions on the issue of global warming.

....

....

....

....

....

....

....

....

Chapter 8: Human Systems and Resource Use

8.1 Human Population Dynamics

8.1.1 Demography

The study of human populations is called 'demography'. The first part of this chapter looks at the demography of human populations in detail. It is recommended that you first understand the population model from Chapter 2 and you learn the key terms in Table 8.1.

Term	Definition	Examples and equations
Crude birth rate (CBR)	Annual number of births per 1000 population	The average CBR of industrialised coun-tries in 1800 was 38 per 1000
Crude death rate (CDR)	Annual number of deaths per 1000 population	The average CDR of non-industrialised countries in 1900 was 40 per 1000
Fertility	Reproductive potential of the population	Fertility can be measured by crude birth rates or the refined indexes described below
Fertility rate	Average number of births per women of child-bearing age range	Average fertility rate in the US
Total fertility rate (TFR)	Average number of children a women has in her lifetime	Total fertility in Niger in 2009 was 7.07 children per women
Doubling time (DT)	The number of years it would take a population to double. For a natural increase rate of 1% this is 70 years	Doubling time = 70/natural increase rate
Natural increase rate (NIR)	Rate of growth in human populations Note: equation is divided by 10 as CBR and CDR out of 1000	% natural increase = (CBR-CDR)/10

Table 8.1: **Key terms in demography**

This 'rule of 70', and a similar 'rule of 72', can be very useful. For the maths behind it, see https://en.wikipedia.org/wiki/Rule_of_72

Remember you should use the rule of 70 for doubling time in the exams.

8. HUMAN SYSTEMS AND RESOURCE USE

8.1.2 The Population Explosion

The rapid (exponential) growth of the population in the last century compared with the slow growth of the rest of human history is called 'The Population Explosion'. Figure 8.1 shows the global population over the last 2000 years. Notice the sudden change in slope at the end of the 20th century. It took until the beginning of the 1800s for the human population to reach 1 billion people, but as population went from 4 to 7 billion each additional billion milestone passed in **less than 13 years**.

There is considerable uncertainty in the models, as shown by the 2 billion variation between the upper and lower margins of the UN predictions for 2050. Much of this uncertainty is due to difficulty in predicting future fertility. This has been seen to vary greatly in the past due to changes in female education, health care and rapid urbanisation.

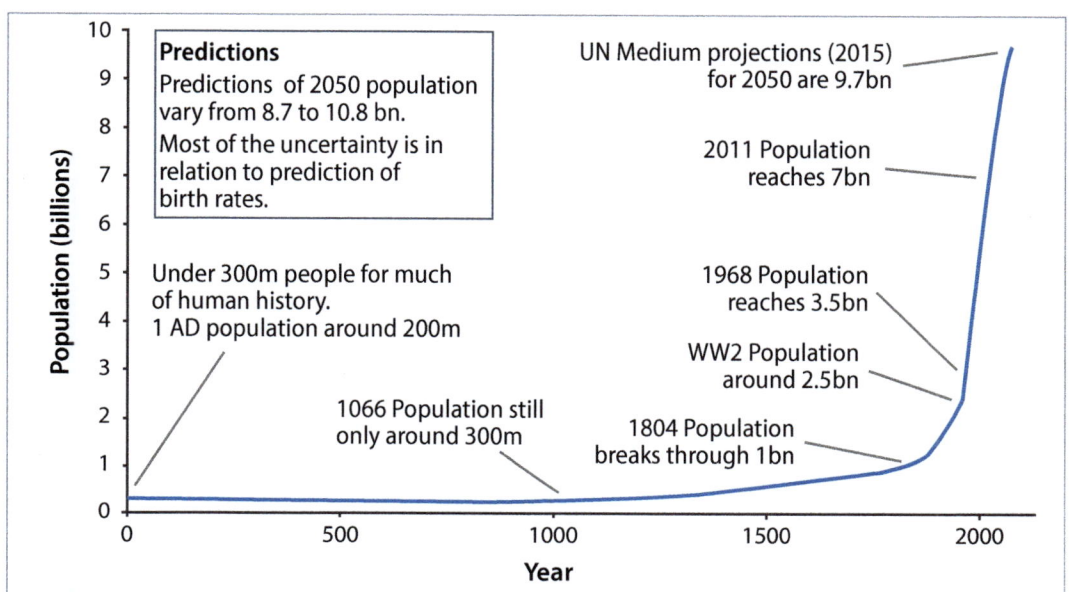

Figure 8.1: **Graph showing world population growth**
Source: Data from United Nations, 2008; and Kunzig, 2011

Continued growth of the human population places increasing pressure on the Earth's life support systems and resources. Population can be seen to drive many environmental problems, but, as discussed later, this should also be considered alongside changes in development and economy.

Global population data can be found at https://www.un.org/development/desa/pd/

Practice questions: Human Population Dynamics

1. *Define* the following terms:

Term	Definition
Crude birth rate (CBR)	
Crude death rate (CDR)	
Fertility	

181

Term	Definition
Fertility rate	
Total fertility rate (TFR)	
Doubling time (DT)	
Natural increase rate (NIR)	

2. Using the data presented on CBR and CDR, *calculate* the missing values of natural increase (%NIR) and complete the table.

Year	% NIR	CBR	CDR
1950–55	1.77	37.2	19.5
1955–60	1.8	35.3	17.3
1960–65	1.94	34.9	15.5
1965–70	2.02	33.4	13.2
1970–75	1.94	30.8	11.4
1975–80	1.77	28.4	10.7
1980–85	1.76	27.9	10.3
1985–90	1.75	27.3	9.7
1990–95	1.54	24.7	9.4
1995–2000	1.36	22.5	8.9
2000–05	1.26	21.2	8.6
2005–10	1.18	20.3	8.5
2010–15		19.4	8.3
2015–20		18.2	8.3
2020–25		16.9	8.3
2025–30		15.8	8.5
2030–35		15	8.8
2035–40		14.5	9.2
2040–45		14	9.6

3. The table in Q2 is the medium range projection from the UN. *Suggest* why there is so much uncertainty in predictions of what the population will be in 2050.

....

....

....

4. *Suggest* which predicted measure you think has the largest uncertainty. *Explain* why this is.

....

....

....

....

....

....

8.1.3 Age-gender Pyramids

There is considerable variation in age structures in populations in relation to level of development, location, and other socio-economic factors. The pyramid in Figure 8.1 illustrates data on the number of individuals in a village in each five-year age class and their gender. For example, there are five males in the village between the ages of 36–40 but only one female between 6 and 10 years old.

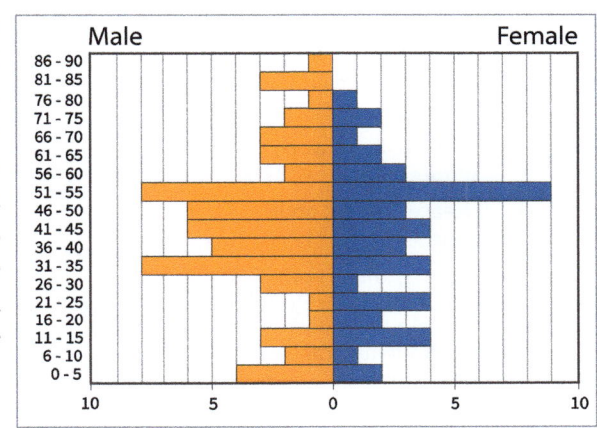

Figure 8.2: **Age-gender pyramid for a small rural village of 108 people**
Source: Tuffin & Sae-Tang, 2008

8.1.4 Demographic Transition Model (DTM)

Demographic transition refers to observed changes in populations and predicts a pattern of decline in mortality and fertility over time due to social and economic development. This model shows what happens to populations over time; the pattern has been shown to be generally true in many countries through their history and in the world as a whole. The model is shown graphically in Figure 8.3 and summarised in Table 8.2.

The model shows that the fall in death rate creates the natural increase, not an increase in birth rate. The biggest natural increase occurs in phase 2 of the model, when birth rates remain much higher than death rates.

Demographic transition is linked closely to the level of development in a country; changes can either be seen in historical data of a country or by comparing countries in different states of development.

Although the model shows a general pattern that holds true there have been differences in rates of change. Transition came later and death rates fell quicker in poorer countries during the 20th century and birth rates stayed high for much longer.

The UN publishes a range of demographic data at https://www.unfpa.org/data/world-population-dashboard

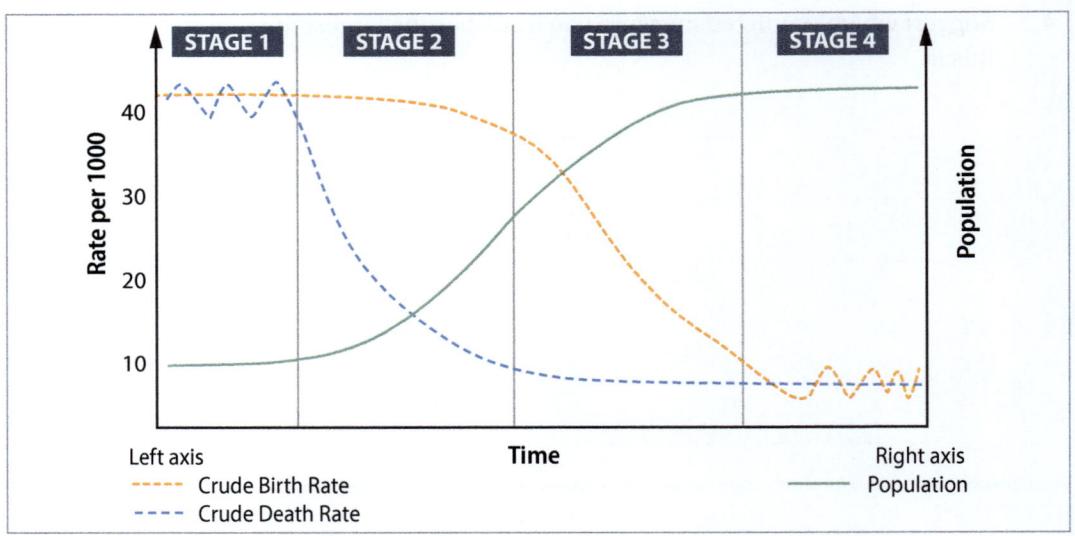

Figure 8.3: **The demographic transition model**
Source: Data from United Nations, 2008; and Kunzig, 2011

Stage	Society	BR	DR	Population	Pyramid shape
Stage One High Stationary	Little or no family planning Large family for work/future security Disease, famine, poor hygiene Hunter gatherers or subsistence agriculture	High, constant	High, fluctuating	Slow growth Pyramid concave E.g. tribal people in the Amazon or Papua New Guinea	
Stage Two Early Expanding	Improved medical care, food security and irrigation Decrease in child mortality Increased life expectancy Improved agriculture, permanent, commercial. Beginnings of industry	High	Falling	Rapid growth Pyramid still wide, but less concave Height grows E.g. Bangladesh	
Stage Three Late Expanding	Family planning improves Perceived need for large families declines Urbanisation leads to falling fertility Industrialisation reduces need for labour	Falling	Low	Rapid growth, slowing towards the end of stage Pyramid becomes convex E.g. Brazil	
Stage Four Low Stationary	Perceived need for small families rises Emancipation of women Increasing value placed on con-sumption Tertiary industry develops (service sector, education, etc.)	Low, fluctuating	Low, stable	High, stable or declining Pyramid becomes inverted E.g., Japan	

Table 8.2: **Summarising the demographic transition**

8.1.5 Factors Influencing Human Population Dynamics

These can be divided into the primary factors of changes in birth rate or death rate, or the secondary factors that ultimately are driving the change in the primary factors. The most significant secondary factors are cultural, historical, religious, social, political, and economic. Urban populations are significantly lower than equivalent rural ones. National and international development policies are also known to have significant effects on population change.

It is useful to explain the impact of secondary factors on the primary; using examples where possible, consider the idea that religious beliefs can maintain high birth rates, for example, Muslim and Christian countries tend to have higher birth rates (other factors being equal). In other words, this is not equivalent to stating that all such countries have higher birth rates as developmental factors can override the influence of religious belief.

8.1.6 Controlling Population Growth

Human population growth is influenced by policies on the international level. Development policies, such as the UN's 17 Sustainable Development Goals, target a number of areas for improvement by 2030. Many of these goals have direct relevance to population growth, for example:

- Reproductive and maternal health (Goal 3)
- Newborn and child health (Goal 3)
- Universal primary and secondary education (Goal 4)
- Gender equality (Goal 5).

These goals illustrate how policy can influence population directly; it has been estimated that fulfilling all of the SDGs by 2030 would lead to a projected population decline by 2100, at between 8.1 and 8.7 billion.

Progress in many of the goals leads to directly reduction in death rates, some indirectly with the same effect. They also influence birth rates. Improving education can lead to reduced birth rates, through raising awareness of choices of birth control and providing opportunities that prolong life in education, so reducing the number of marriages and families starting at a young age. This is particularly true when the education is of women and helps to empower women in society. This has been shown to be particularly effective in Bangladesh and sub-Saharan Africa.

National development policies can also exert significant influence on population dynamics. Sometimes this is a stated objective of the development, such as the Population and Community Development Association in Thailand that has aimed, since 1974, to promote family planning alongside community development and empowering the poor.

Unplanned development can also influence population dynamics. Rapid urbanisation in LEDCs led, eventually, to a lowering of fertility rates and to large adjustments on projections in the 1990s, combined with death rate increases due to HIV/AIDS.

Cultural values are a powerful force in determining population growth. One reason why birth rate decline lags behind death rate decline in the DTM is that cultural attitudes to fertility are slow to change. Changes in death rate are of course easily accepted. Some arguments suggest that religious beliefs, acceptance of materialism and other fundamental values are the most significant in long term population growth.

ENVIRONMENTAL SYSTEMS AND SOCIETIES SL

8.2 Resource Use in Society

8.2.1 Natural Capital

Look back to section 1.4 to review this topic in relation to sustainability and sustainable development.

Natural resources taken from living or non-living systems for human use can be referred to as natural capital. These resources are naturally occurring goods and services. Goods are physical commodities and services are activities that are used as they are produced. Natural capital can be classified into two categories, as shown in Table 8.3.

Natural Capital Class	Examples of Goods	Examples of Services	Ability to regenerate
Renewable (produced by an ecosystem or part of a natural material cycle)	• Food, timber, wool, and many other natural resources • Water stores, groundwater or surface water • Soil nitrate fixed by lightning	• Stable climate, air and water quality, crop pollination and flood protection • Stratospheric ozone layer	Regenerates using solar energy (within human timescales)
Non-renewable (use reduces amount available to society)	• Fossil fuels and minerals like tin ore, formed over millions of years	• Stability of land surface • Supply of soil nutrient from weathered rock	Fixed amount (regenerates within geological time-scales)

Table 8.3: **Two classes of natural capital**

8.2.2 Natural Income

Renewable forms of natural capital can produce a natural income in the form of goods or services, such as those listed above. This is can be seen as similar to interest on capital savings. Natural income can be calculated as sustainable yield. The goods or services may translate to a direct market place value, for example, timber yield; others may be viewed by society as a free resource, for example, absorbing pollution.

8.2.3 Sustainability of Resource Use

Growth
Sustainable yield is equivalent to what we generally refer to as growth of either biomass, energy or population.

Sustainability means living within the natural income without reducing the natural capital. Sustainable yield (SY) is the rate of increase in the natural capital. There are two ways of calculating this for renewable resources.

The first relates to productivity in terms of biomass or energy:

Sustainable Yield (SY) = (total biomass or energy at $t+1$) − (total biomass or energy at t)

In terms of ecosystem productivity, we can see that this calculation of sustainable yield would be equivalent to net primary productivity for a producer yielding timber or grain. It is also equivalent to net secondary productivity. The calculation of sustainable yield for fisheries is explained in section 4.3.7.

Note that other factors need to be considered in terms of impact on the natural income, including pollution, climate change, and other environmental damage caused during extraction or processing of the resource. These factors may reduce productivity and therefore income.

8.2.4 Value of Resources

Natural resources do not only have utilitarian economic and environmental value for goods and services, they have other values to humans. Look back at section 3.4.1 to see some of these detailed. Values can be aesthetic, cultural, ethical, spiritual, technological, social, and intrinsic. Intrinsic values are unrelated to human beings altogether, the concept of intrinsic means within the thing itself. You could consider it as what value the species may place on its own existence.

8.2.5 The Dynamic Nature of a Resource

The status of resources in terms of their usefulness varies in relation to the technological development status of countries, what is valued as natural capital is dynamic. This is shown clearly when new markets evolve creating demand for new resources in specific countries. For example, the elements tantalum (transition metal) and neodymium (rare earth) are both now valuable resources as they are used in portable electronic devices. They were not valuable before the demand increase for mobile electronic devices, and in the future as technology creates demand for different components this value may well change. Prior to the 1800s, lamps and candles were largely fuelled by organic oil, often from the sperm whale. The replacement of this fuel with cheaper crude oil, in the form of kerosene, helped to save the sperm whale from being hunted to extinction as populations were hunted above sustainable yield levels.

Other factors that lead to the status of a resource in a given society include socio-cultural factors to do with religions and traditions. Political factors and trading agreements are also both key factors in determining resource value in different countries across the globe.

8.2.6 Mismanaging Natural Capital

You need to be able to give one example each of how renewable and non-renewable natural capital can be badly managed. There are several examples in this guide already that can be used. Here are two more that both demonstrate the principles on small islands.

Phosphate mining on the island republic of Nauru

Non-renewable resources, such as mining fossil fuels or mineral resources can never be sustainable by definition. However, it can be mined at a carefully considered rate and the habitats restored following mining. On the pacific island of Nauru, phosphate was strip mined from over 80% of the island area causing environmental degradation and extinction of indigenous species. Little mining can now take place and there is a lack of land for farming and housing.

Wood harvest from Easter Island

Renewable resources can be managed at a sustainable rate. Easter Island is a Pacific island famous for the disappearance of its people. When the first European explorers arrived (1700s), the island was populated; by the late 19th century most of the people had gone dropping to a low of only 111 people in 1871. The forested island had been cleared of its trees, which meant that the society was no longer able to build boats and fish on the ocean. It is believed that this contributed to the decline of the human population there.

ENVIRONMENTAL SYSTEMS AND SOCIETIES SL

Practice questions: Resource Use in Society

1. *Calculate* doubling time to complete the following table.

Year	% Natural Increase Rate	Doubling Time (Years)
1950–55	1.77	40
1955–60	1.8	39
1960–65	1.94	36
1965–70	2.02	35
1970–75	1.94	
1975–80	1.77	
1980–85	1.76	
1985–90	1.75	
1990–95	1.54	
1995–2000	1.36	
2000–05	1.26	
2005–10	1.18	
2010–15		
2015–20		
2020–25		
2025–30		
2030–35		
2035–40		
2040–45		

2. *State* which decade showed the shortest doubling time?

....

3. Referring to the pyramid in Figure 8.2 on page 183, *state* the oldest possible age of a man and a woman in the village.

....

4. *State* the number of females aged 51–55 in the village.

....

5. *Outline* the demographic transition model.

....

....

....

6. *Explain* the changes in global doubling time calculated above in relation to demographic transition.

   ....
   ....
   ....

7. *State* which stage of the DTM has the greatest natural increase.

   ....

8. *State* which factor initially changes to lead to this increase in population.

   ....

9. *State* which stage of the demographic transition has a concave pyramid.

   ....

10. *Evaluate* the application of the demographic transition model to future populations.

    ....
    ....
    ....
    ....

11. *Define* natural capital and its two forms.

    ....
    ....
    ....

12. *Define* the term 'sustainability'.

    ....
    ....
    ....
    ....

13. *Explain* the concept of sustainability in relation to the idea of natural capital and its income.

    ....
    ....
    ....
    ....

14. Complete the following table:

Natural Capital Class	Examples of Goods	Examples of Services	Ability to regenerate
	Food	Stratospheric ozone layer	
		Supply of soil nutrient from weathered rock	

15. The way that society needs and values a resource may change. **Outline** how this supports the argument that even if resources are used up it is sustainable, as new resources are likely to be found to replace them.

....

....

....

....

16. **Discuss** why in a debate regarding commercial whaling, it is hard to quantify the intrinsic value of the blue whale and to compare this with the commercial value of its flesh?

....

....

....

....

17. Blue whales (*Balaenoptera musculus*) populations in the North Atlantic along the east coast of North America are currently estimated at around 500 individuals, fewer than 5% of the original population before commercial whaling began. In one year the population increases by 20 individuals. **Calculate** the sustainable yield.

....

....

8. HUMAN SYSTEMS AND RESOURCE USE

18. Coppice trees are harvested by cutting back to the roots and allowing small shoots to grow. Trees are cut on a cycle of around 10–15 years, depending on the species and the use of the wood. In a coppiced woodland the biomass in one year is found to be 14 tonnes/ha, the following year it is 19 tonnes/ha. **Calculate** the sustainable yield.

....

....

8.3 Solid Domestic Waste

8.3.1 Types of Solid Domestic Waste

Solid domestic or municipal waste (rubbish or garbage) from residential housing includes a mixture of the following:

- Paper (20–30%)
- Glass (5–10%)
- Metals (5%)
- Plastics (5–10%)
- Organic waste from kitchen or garden (20–50%)
- Potentially hazardous materials, for example, electronic items, batteries, medicines, pesticides, cleaning products etc. (5–15%).

Faeces
Human faeces is *not* an example of solid domestic waste.

Approximate percentages give a guide to volumes for a typical MEDC. Total mass of waste produced varies between developed countries, but it can be over 800 kg per capita per year. The composition of solid domestic waste varies considerably. Plastics, electronic waste, and batteries have become a particular concern as they are non-biodegradable and often contain toxic materials.

8.3.2 Strategies for Solid Municipal Waste Management

Level	Strategy	
Cause	Reduce	Reduce use of excess packaging that cannot be reused
	Reuse	Within the home and through exchange of second hand materials
		Choose reusable containers such as glass milk bottles instead of plastic
	Recycle	Choose materials that can be recycled
	Composting	Organic waste can be composted at source
		Worm bins or compost heaps can compost most vegetable waste
		Choose biodegradable materials which can be composted.
Release and transfer	Separate wastes	Separation of domestic waste into categories of paper, plastic, glass, metal, and organic can help reuse or recycling schemes Compulsory in some countries
	Legislation	Laws and taxes to encourage recycling and waste separation

Level	Strategy	
Effects	Incineration	• Incinerators burn rubbish • If it is not carefully sorted before combustion, then high levels of toxic pollution may enter the atmosphere • Waste cinders, which are often toxic, need disposal
	Landfill sites	• Rainwater leaches through landfill sites and carries dissolved pollutants into the groundwater (see Love Canal case study)
	Sealed landfill site	• Sealing prevents water from entering the rubbish and taking the pollution out; this can prevent toxins from entering the drinking water supply or natural ecosystems
	Waste to energy schemes	• Using waste as fuel, to generate biogas
	Cleaning up waste	• Reclaiming landfill • Removing waste from ocean, e.g., Great Pacific Garbage Patch
	Composting	• Composting schemes can turn organic waste into a resource which can be sold back for use in gardens

Table 8.4: **Summary of waste management strategies**

8.3.3 Evaluating Waste Management Strategies

Suitable strategies will vary depending on the location and the type of waste available. Landfills can be designed to use the waste gas methane for power production. Disposal of e-waste is a growing concern due to both pollution and the waste of resources. Regulations regarding design and disposal of electronic goods in some countries are starting to help reduce the impact. The most sustainable schemes will turn waste into resources with minimal environmental impact, breaking the linear resource pathway and developing a circular economy. Consider the environmental impacts of these two different models shown in Figure 8.4.

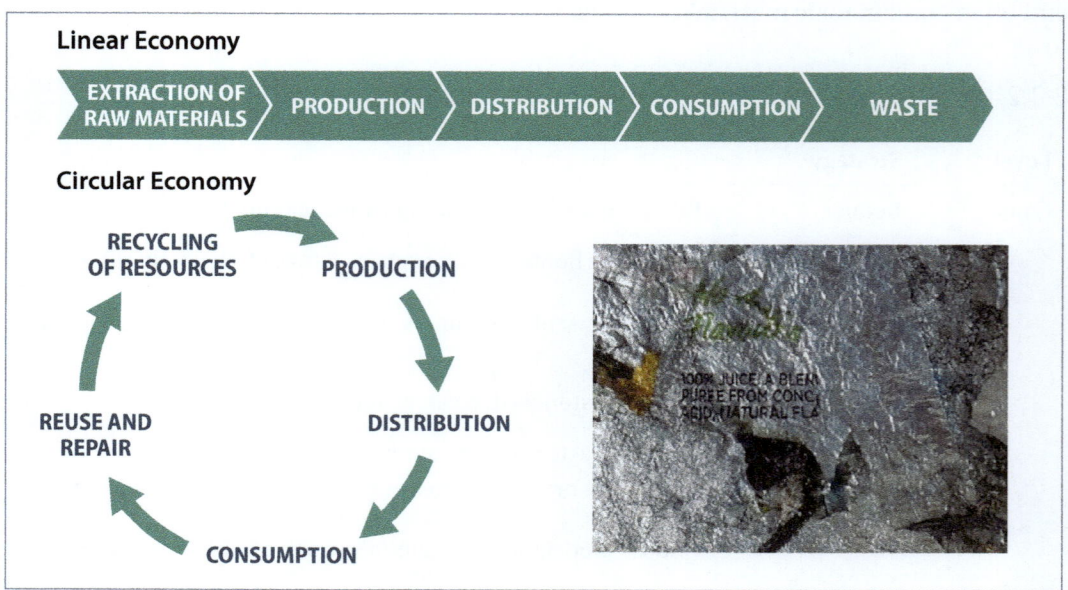

Figure 8.4: **Contrasting the linear and circular economic models**
(Inset: 'ecoboard panel' from author's house—insulating wall panel made from drink cartons)

8. HUMAN SYSTEMS AND RESOURCE USE

8.3.4 Human Factors Influencing Choice of Waste Management

Cultural, political, and economic factors all influence the choice of management strategy and its effective implementation.

Cultural factors are reflected in the dominant attitudes towards the environment and in the accepted ethics of the society. If society adopts an un-enquiring 'out of sight, out of mind' approach, then it may be easy for individuals to dispose of pollution in a hazardous way.

Political factors set the legal framework for regulation or governments may be in control of the pollution directly. Governments vary in their openness towards permitting public identification of pollution sources and management.

Economic factors can promote pollution if the environment is seen as a free resource, external to the market place. This allows for apparently free waste disposal.

Consider these factors in relation the Love Canal case study below.

8.3.5 Love Canal Case Study

Love Canal was the name given to an unfinished canal designed by William Love and dug in the town of Niagara Falls, USA. Funding to dig the canal ran out and, in the 1920s, it was turned into a landfill site—a dumping ground for domestic and chemical wastes from industry and manufacturing. The landfill site was poorly managed and later covered with earth. It was sold for only one dollar in the 1950s. The land was then used to build over 100 homes and a school.

Chemical cocktail
Over 100 different chemical compounds were identified at Love Canal. At least 11 of them were human carcinogens.

Residents complained of unpleasant smells at first. By the late 1960s chemical waste was seeping into the school basement and by the 1970s some of the corroding waste disposal drums could be seen in gardens. Patches of grass and trees were found dying. Many medical cases were later linked to the pollution, including cases of birth defects, miscarriage and skin burns. The chemical mixture found in the landfill included high levels of the chemicals benzene and polychlorinated biphenyls (PCBs), both known to be human carcinogens.

In 1978 the area was declared as in a state of emergency by the President and emergency financial aid was approved (previously only used for natural disasters). 900 families were evacuated from the area. When the land was sold in the 1950s it included a warning regarding chemical waste on the property, yet the decision was made to build a school on the land. The people who bought houses at Love Canal were not informed of this danger. Children who played in the area were not aware that the pools of chemical waste should be avoided. In the 1990s the chemical company responsible for managing the original landfill agreed to pay over 225 million US dollars for the clean-up costs and to compensate victims (Centre for Health, Environment and Justice).

8.4 Human Population Carrying Capacity

8.4.1 Carrying Capacity

Carrying capacity is defined as the maximum number of people that can be sustainably supported by a given environment. This figure is elusive to even experts as the total number includes a prediction of future behaviour, technological developments and the dynamic resource demands by populations.

Table 8.5: **Estimates of global carrying capacity**
Source: From Cohen (1995)

	Global Carrying Capacity (Billions of People)	No. of Estimates Made
Pre 1750	6.5–10.5	3
1751–1900	6–10.6	2
1901–50	2–150	17
1951–2000	1–1000	42

Estimates of carrying capacity vary considerably and even though over time more estimates are made there is less certainty on the actual number. Table 8.5 shows estimates varying between 1 billion and 1,000 billion people for global carrying capacity.

Why should there be so much difficulty in calculating true global carrying capacity? Most estimates are based on the amount of food that can be sustainably produced as a significant limiting factor. Importantly, this also involves predicting future resource issues in relation to food, energy and water. Cohen (2010) named three schools of thought to increase global carrying capacity—see box.

Cohen's three schools model

Bigger pie: new technology to increase productivity

Fewer forks: reduce consumption

Better manners: better government and fairer markets.

8.4.2 Local Carrying Capacity

Carrying capacity estimates must take into account the role of technology in extending productivity and also variability in standards of living accepted both culturally and individually.

Local estimates of carrying capacity are also hard to establish as international trade can overcome local limits to growth. When London first emerged as a city of 1 million in the 1800s, it was mostly sustained by farms within England. In this century, lamb eaten in London is likely to have been farmed in New Zealand.

8.4.3 Increasing Carrying Capacity

To increase carrying capacity import of materials or energy to an area is not a solution. Technology can only improve efficiency or productivity but may require more inputs of energy and materials.

There are methods available to increase the carrying capacity of land. Improvements in agricultural productivity increase carrying capacity; however, if these systems are fossil fuel dependent, they are not sustainable. Land use systems that improve productivity and use the natural regeneration of the agricultural system increase carrying capacity. Agricultural technologies that use renewable resources to improve productivity can improve carrying capacity, such as solar powered irrigation systems.

Carrying capacity is also determined by the rates of resource use and energy use by society, combined with impacts on life support systems through pollution or habitat destruction. Table 8.6 shows schemes for actively increasing carrying capacity.

	How?	Effect
Energy conservation	Improving efficiency of any energy demanding system, such as housing design, to reduce heat loss or gain	Reduces total amount of electricity that needs to be produced by power stations
Reduce, reuse, recycle	These three ways to lower demand for material goods can increase carrying capacity	Reduces total demand for a given resource but may have unwanted effects from pollution (recycling)

Table 8.6: **Methods to increase carrying capacity**

8. HUMAN SYSTEMS AND RESOURCE USE

8.4.4 Ecological Footprints

The ecological footprint was introduced in section 1.4.8. It is the total area of land required to supply our needs and absorb our wastes per capita.

Mathematically, the footprint is the inverse of the carrying capacity:

$$\text{Ecological footprint (ha/person)} = \frac{1}{\text{Carrying capacity (people/ha)}}$$

Although, mathematically, the carrying capacity can be calculated from the footprint, it cannot be used to find out planetary carrying capacity as this involves a prediction of future behaviour. Hence, the number produced would be correct assuming no change in the future population in terms of development and consumption, which is an inaccurate projection.

Natural regeneration and, therefore, ability to absorb pollution is also not constant between countries due to variations in primary productivity of both natural biomes and agricultural. Consequently, land in high latitudes has a lower carrying capacity for either resource supply or waste removal.

8.4.5 A Simple Ecological Footprint Calculation

For an assessment of ecological footprint, the following two equations can be used:

1. Food resource supply equation

$$\text{Per capita land requirement for food production (ha)} = \frac{\text{Per capita food consumption (kg/yr)}}{\text{Mean food production per hectare of local arable land (kg/ha/yr)}}$$

2. Pollution absorption equation

$$\text{Per capita land requirement for absorbing waste } CO_2 \text{ from fossil fuel (ha)} = \frac{\text{Per capita } CO_2 \text{ emissions (kg C/yr)}}{\text{Net carbon fixation per hectare (kg C/ha/yr)}}$$

> **Ecological footprints**
> This is a worked example to deepen your understanding. You don't have to learn the equations, but you should understand the principle and be able to compare two countries.

If these two figures are calculated, then the sum of them is the ecological footprint per capita, which gives a simplified assessment of the ecological footprint of a country. This method ignores other elements commonly included in more complex footprint calculations such as:

- Food production from aquatic systems
- Production of other resources or ecosystem services from land
- Production and absorption of wastes other than CO_2
- Land taken up by housing and infrastructure.

8.4.6 Comparisons of the Ecological Footprint (EF) from an LEDC and a MEDC

Brazil is given as an example here of an LEDC and Canada as a MEDC. The data is from Metcalfe, (2010) except for carbon figures which come from Begon, Harper, and Townsend (2006). A simple footprint can be calculated using the equations in section 8.4.5.

	Per Capita Food (Grain) Consumption (kg/yr)	Mean Food (Grain) Production per Hectare of Local Arable Land (kg/ha/yr)	Per Capita CO_2 Emissions (kg C/ha/yr)	Net Carbon Fixation per Hectare (kg C/ha/yr)
Brazil	210	2,919	1,900	10,000
Canada	710	3,031	16,900	4,000

Figure 8.5: **Raw data for footprint calculations for Brazil and Canada**
Source: Data from Metcalfe, 2010; and Begon, Harper, and Townsend, 2006

$$\text{Brazil EF} = \frac{210}{2,915} + \frac{1,900}{10,000} = 0.072 + 0.19 = 0.26 \text{ ha}$$

$$\text{Canada EF} = \frac{710}{3,031} + \frac{16,900}{4,000} = 0.23 + 4.25 = 4.45 \text{ ha}$$

It should be noted that these are much smaller footprints due to simplification of the calculations, excluding the factors bullet pointed above in section 8.4.5.

There is a clear difference between these two countries. Food consumption is based on per capita use of grain. In MEDCs much of this is fed to livestock for meat production, which leads directly to a much higher total grain consumption. Food production per ha is slightly higher in Canada, reflecting the intensive agriculture. In Brazil, agricultural productivity is also fairly high due to the conditions for crop growth. Carbon dioxide emissions per capita are far greater in an MEDC than an LEDC. The tropical rainforest biome type and other biomes in the tropics combine to give a much higher carbon fixation rate than the colder tundra, forests, and grasslands of Canada.

Practice questions: Solid domestic waste/Human population carrying capacity

1. Using examples, how do cultural, economic and political factors influence how societies manage pollution?

Cultural

Economic

Political

2. *List* the types of solid waste that are produced by you and your community.

....

....

....

....

3. *Describe* management strategies for municipal waste using a table.

4. *Evaluate* the strategies you have given.

....

....

....

5. *Explain* the difficulty in applying the concept of carrying capacity to human populations.

....

....

....

6. *Describe* and *explain* ways in which a human population can increase carrying capacity through changing how resources and energy are managed.

....

....

....

7. *Define* the term ecological footprint and explain how it relates to carrying capacity.

8. Look at the data in the table below:

	Per Capita Food (Grain) Consumption (kg/yr)	Mean Food (Grain) Production per Hectare of Local Arable Land (kg/ha/yr)	Per Capita CO_2 Emissions (kg C/ha/yr)	Net Carbon Fixation per Hectare (kg C/ha/yr)
Country A	330	4,500	9,900	10,000
Country B	750	6,000	19,000	3,000
Country C	240	3,500	970	6,000

State which country is more likely to be:
(a) An MEDC

(b) In a tropical region

(c) A country in the early stages of the DTM

9. *Discuss* how national and international development policies impact on human population growth.

8. HUMAN SYSTEMS AND RESOURCE USE

10. ***Discuss*** how much influence culture has on human population growth.

11. ***Compare*** the ecological footprints of two countries, one MEDC and one LEDC. ***List*** reasons for the differences in the size of the footprints.

ENVIRONMENTAL SYSTEMS AND SOCIETIES SL

Answers to practice questions

Chapter 1

Environmental Value Systems

1. This is the 'world view' or set of paradigms that shape the ways that individuals and groups approach environmental issues.

2. The model is useful as a tool for analysis as it simplifies situations which can allow for patterns to be revealed and hypotheses suggested. For example, students that have studied ESS are more likely to adopt personal lifestyle change than those that haven't. Clearly though there are great simplifications with using such models, and the way people think and act is likely to be considerable more complex. Overly relying on models may lead to stereotypical views or polarised views of debates.

3.

Environmental Philosophy	Outline of key ideas, people or actions
Deep Ecologists	A need for spiritual revolution to fix environmental problems is at the core of all environmental issues. Nature at the centre, equal rights for species. Activists such as Earth First!
Soft Ecologists	Self-sufficiency in resource management. Ecological understanding is a principle for all aspects of living, e.g. permaculture.
Environmental Managers	Promote working to create change within the existing social and political structures. Current economic growth can be sustained if environmental issues are managed by legal means or political agreement, e.g. Government and NGOs.
Cornucopians	Environmental issues not really 'problems', as humans have always found a way out of difficulties in the past. New resources and technologies will solve any environmental issues as they are encountered. There is no need for radical agendas, socio-economic or political reform, e.g. European settlers or free-market economists.

4. See timeline examples in Table 1.4.

ANSWERS TO REVISION QUESTIONS

5. (a) Human population growth

Deep Ecologist	Reduce population growth through personal actions.
Soft Ecologist	Reduce population growth through personal actions.
Environmental Manager	Intervention and family planning through government action would be the best solution.
Cornucopian	Don't attempt to control, allow birth rates to be controlled by costs of having children.

(b) Global food supply

Deep Ecologist	Reduce food demands, include some aspects of SE.
Soft Ecologist	Provide food to local population through sustainable agricultural systems, such as permaculture.
Environmental Manager	Promote changes in existing food supply system to encourage sustainable practices.
Cornucopian	Allow market-based mechanisms to control supply of food through changes in the demand. Changes in agriculture will cover any shortfall in production.

(c) Energy supply choices

Deep Ecologist	Reduce demand through lifestyle change.
Soft Ecologist	Reduce demand and produce more power on a small scale.
Environmental Manager	Convert the grid system to supply power needs via alternatives or nuclear (much debate about nuclear as an environmental choice).
Cornucopian	Allow market to make choices on the supply of energy.

(d) Water resource management

Deep Ecologist	Use less water at source, reduce consumption, e.g. low water usage toilets and showers.
Soft Ecologist	Provide local solutions to use water more efficiently, e.g. trickle drip irrigation.
Environmental Manager	Adjust the water supply system to provide water more efficiently or from new sources, e.g. desalination.
Cornucopian	Allow markets for private supply of water to distribute efficiently.

(e) Conservation of biodiversity

Deep Ecologist	Value species equally, protect all wildlife and animal rights.
Soft Ecologist	Greater emphasis than EM on valuing many species.
Environmental Manager	Protect endangered species and significant areas.
Cornucopian	Protection is sentimental, extinction is natural. No need to protect biodiversity, if it is of value people will look after it.

(f) Management of a named conservation area, e.g. Snowdonia National Park

Deep Ecologist	Wildlife has intrinsic value and should be left alone. Likely to support rewilding.
Soft Ecologist	Habitat management may include cutting and clearing, reintroduction of species.
Environmental Manager	Habitat management should be applied to maintain maximum biodiversity, e.g. coppice cut rotation in woodlands.
Cornucopian	Allow the change to take place, and don't bother to manage. Wildlife should be exploited through hunting and gathering to generate money. Nature will restore the capital over time.

(g) Ecological footprints

Deep Ecologist	Reduce footprints by tackling individual consumption.
Soft Ecologist	Reduce footprints by promoting self sufficiency and bioregionalism.
Environmental Manager	Encourage sustainble practices into the city through legislation and policy.
Cornucopian	Ecological footprints don't need to be reduced, sustainability problems will be solved through market mechanism.

(h) Love Canal

Deep Ecologist	Do not create the initial demand for the products that generate the waste.
Soft Ecologist	Manage through landfill legislation and controls on waste disposal.
Environmental Manager	Manage through landfill legislation and controls on waste disposal.
Cornucopian	Companies should not be restricted by legislation, damaged land will be devalued and people will relocate.

(i) Eutrophication

Deep Ecologist	Reduce sources of phosphate and nitrate, e.g. use phosphate free detergents, reduce waste.
Soft Ecologist	Use organic fertiliser and promote organic farming.
Environmental Manager	Use buffer strips or direct habitat management techniques, e.g. reed bed filters to reduce impacts.
Cornucopian	Allow the change to take place, it is natural and new habitats will develop over time. There are other lakes.

(j) Photochemical smogs

Deep Ecologist	Reduce air pollution through lifestyle change, e.g. drive less.
Soft Ecologist	Reduce air pollution through lifestyle change, e.g. drive less.
Environmental Manager	Manage air pollution through local laws in the city.
Cornucopian	Don't manage the pollution, allow markets to operate freely. People can choose where to live.

ANSWERS TO REVISION QUESTIONS

(k) Depletion of stratospheric ozone

Deep Ecologist	Reduce production of ODS by lifestyle change, e.g. don't use aerosols.
Soft Ecologist	Reduce production of ODS by lifestyle change, e.g. don't use aerosols.
Environmental Manager	Manage through international protocols, e.g. Montreal Protocol.
Cornucopian	The problem does not require management, the cycle of ozone depletion may not be linked to human emissions anyway. (NOTE: this argument was only stated in the early days of the ozone issue.)

(l) Acid deposition

Deep Ecologist	Reduce demand for energy, e.g. from coal fire power stations.
Soft Ecologist	Reduce demand for energy, e.g. from coal fire power stations.
Environmental Manager	Reduce acidity in emissions by using limestone scrubbers in the power station.
Cornucopian	The problems of acid rain will be resolved naturally. Degraded habitat will return.

(m) Sustainable development

Deep Ecologist	The problem must be resolved by reconnecting at a spiritual level to the biosphere in order to reduce consumption.
Soft Ecologist	Through lifestyle change and sustainable agriculture, reduced consumption levels will lead to sustainability.
Environmental Manager	Sustainable development can be achieved through international regulation and government action.
Cornucopian	Sustainable development is natural, there is no need for international government action, the free market will find the resources for society.

6. Intrinsic values are fuzzy, in other words, hard to clearly define or quantify. This also means that economists struggle to award these financial values. Most people accept that there is some kind of intrinsic value to existence for non-human species and environments. Consider removing all of Antarctica from the world. Supposing it had no impact on the rest of the planet (which of course it would), and you have no direct experience of its existence or non-existence, would this matter? This is your intrinsic value to the existence of Antarctica.

7.

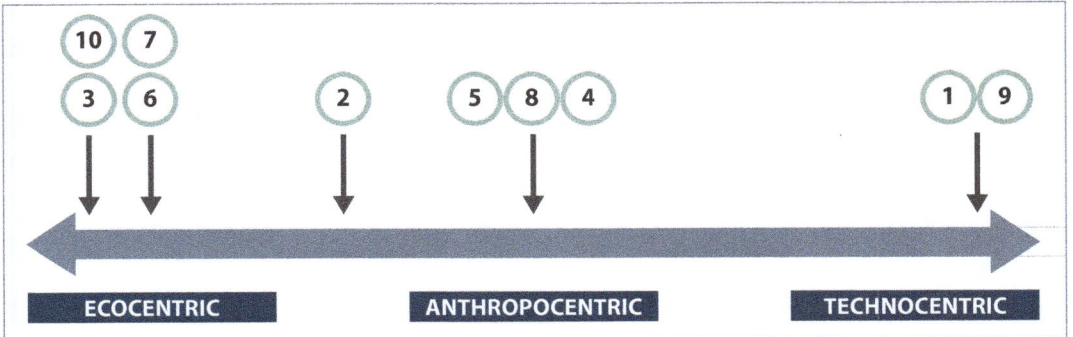

8. For example, I agree with the three quotes that are closest to the anthropocentric viewpoint on the scale above: 5, 8, and 4. These three are closest to my view that the only solutions to attain sustainability are political and will focus on societies' economic and government structures. Whilst changing values and attitudes to reduce consumption are important alone, this approach will not reach enough people and cause the required changes. Sustainable management of the planet will require governments and intergovernmental organisations to guide people and society to more sustainable pathways for energy and resource use.

Systems and Models

1. A system is defined as having two or more interactive parts; connections can be via simple interaction or more complex feedback loops. A system often shows a character of complexity that is more than the sum of its parts, called an emergent property. See the main text for examples.

2. **Open systems**: Human body, freshwater pond, temperate forest.

   Closed systems: Sealed aquarium tank, Winogradsky column, Biosphere 2.

   Isolated: Entire universe, a vacuum sealed box, core of a nuclear fusion reactor.

3. **Matter** is anything that takes up space and has mass. It is normally made of atoms.

   Energy is the ability to do work and affect either the transformation (change) or transfer (movement) of matter.

4. A model is a simplified version of reality used to make predictions or analyse situations. Examples include the Demographic Transition Model, the Niche Model, and the Daisyworld Model. Models can be represented as diagrams, flow charts, or systems diagrams.

Energy and Equilibria

1. The total energy of the universe or any isolated part of it will be the same before and after matter is moved or transformed.

2. In an isolated system the total amount of entropy (disorder) will tend to increase.

3. The first law states that energy is neither created nor destroyed in an isolated system.

   The second law states that an isolated system will become more disordered over time. The entropy law is relevant to environmental systems as it describes how energy conversions are not entirely efficient: some energy is always dispersed to the environment. However, the Earth's environmental systems are open rather than isolated. They import energy due to the constant input of the sun and use this to work against the entropy law. Many environmental systems are, therefore, resilient to change and remain stable. However, if the sun's input were to fail, then the planet would start to obey the second law, increasing in entropy.

4. $168\,J + 12\,J = 180\,J$.

   (J = joules of energy. Note: a joule is the amount of energy transferred when one Newton of force acts on an object in the direction of the force over one metre, or the heat required to raise the temperature of 1 g of water by 0.24°C.)

5. In a sealed box without energy, the plant cannot carry out photosynthesis, so due to the first law its energy content is fixed. The plant needs to respire to survive, using its fixed store of chemical energy which is lost as heat due to the second law. The plant biomass and energy decline, leading eventually to death and disintegration showing the higher entropy of the second law.

6. Negative feedback leads to a greater stability.

7. (a) Negative; (b) Positive; (c) Negative

ANSWERS TO REVISION QUESTIONS

8. Human body temperature regulated by negative feedback loops (note: details of processes, e.g. sweating to cool down or increase in respiration to warm up could be added). Note the equilibrium is maintained.

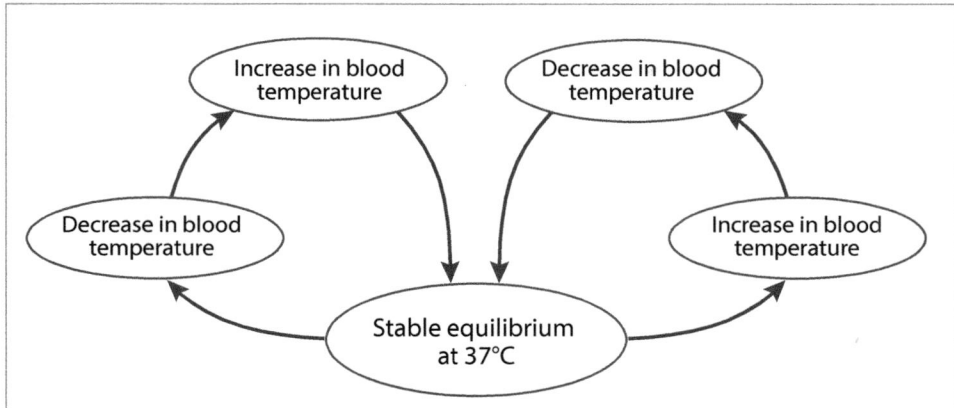

Arctic Ocean temperature changes show positive feedback loops. Currently we are on the right-hand side of this change; during an ice age cooling would have produced the left-hand side changes. Note the changing equilibrium. Arctic Ocean temperature changes showing positive feedback loops. Currently we are on the right-hand side of this change; during an ice age cooling would have produced the left-hand side changes. Note the changing equilibrium.

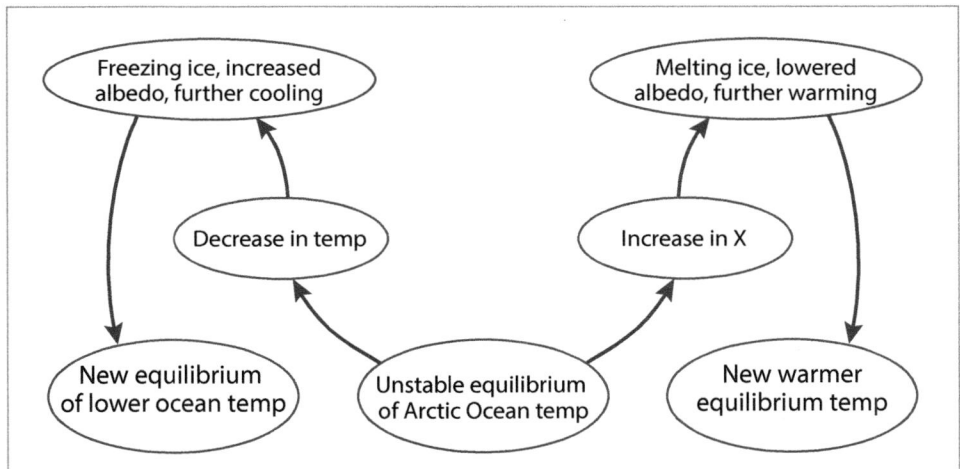

9. In a systems diagram, flows are shown with arrows and units must have time to show rate. Stores are shown with boxes; units do not need time.

10. Total Inputs: 198.5 Gt per year (Fossil fuel and cement [6.2], respiration and decomposition [101], and biological and chemical processes to ocean [90.8]). Total outputs: 193.9 Gt per year (GPP Global [101.5], and biological and chemical processes from ocean [92.4])

    The change in store = input − output

    198.5 − 193.9 = net gain of 4.6 Gt per year

11. (a) No; (b) Yes; (c) No

12.

What are the limitations of knowledge?	Estimations of carbon in a forest are very difficult, even as a standing storage. E.g. carbon is stored in fine roots, bacteria, dead organic matter, trees, leaf litter etc. Furthermore, rates of change of carbon are unpredictable, e.g. forest fires can release carbon rapidly.
How reliable or representative is the data?	There is a very large uncertainty figure for storage amounts, as this data is only available based on a few examples and local variation is great. Small changes in soil and microclimate produce large changes in the forest.
Is it an evidence based model that makes logical sense?	The models are based on limited evidence, for human emissions there is a lot of data, but for ecosystems storage data is more limited.
Are there feedbacks, with possible tipping points?	Yes, these are present at many levels in the model.
Are there possible influences on the model that are missing?	Yes. Variation is far more complex than shown in this simplified model. For example, Land Biota refers to all terrestrial biomes, with variation from savanna to tundra.
Is the model a useful tool for analysing the situation?	Carbon cycles, even though they are an incomplete model, are a useful tool for analytical understanding of the whole system.

13. Daisyworld feedback loop. Note: compare this with the response of the non-living ice ocean system shown in the answer to question 3.

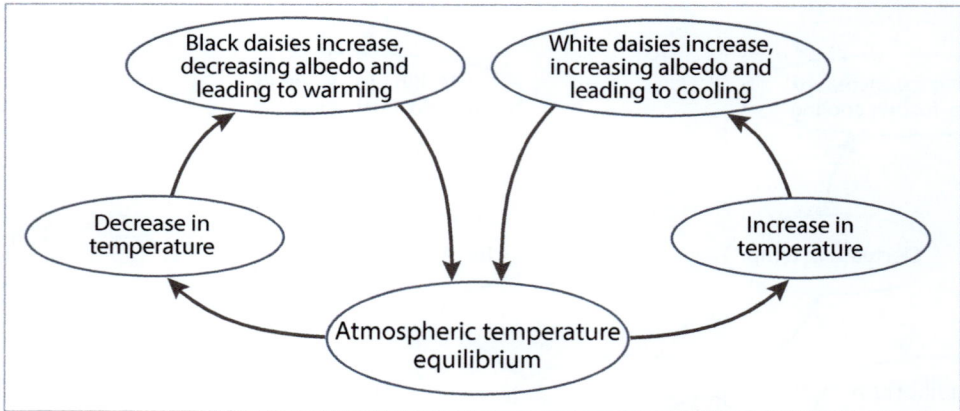

14. The advantages of this model are that it allows for improved understanding of emergent properties in the system, and that is simple enough to give a good conceptual understanding of the system behaviour. Disadvantages are that the model is not nearly complex enough to represent the behaviour of the global ecosystem. Many other factors that influence temperature are not included.

15. Small-scale systems can be seen in the metabolic pathways of living things, such as respiration, showing an interacting system of chemicals and reactions. On a larger scale the living organisms show systems of internal organs that interact. These organs work together to a common purpose showing interactions and emergent properties; examples include food being fully digested in the digestive systems or oxygen reaching the cells via the circulatory system.

Sustainability / Humans and Pollution

1. Roosevelt's quote illustrates the concept of sustainability, as this proposes we live within the natural income of our natural capital and not undermine the capital itself. If the natural capital is undermined by harvesting the resources beyond sustainable yield, then the capital and yield

ANSWERS TO REVISION QUESTIONS

(income) will both be reduced. As Roosevelt says it is better to increase rather than decrease our natural productivity.

2. Ecosystem services include: stable climate; air and water quality; crop pollination; flood protection; stability of land surface; supply of soil nutrient from weathered rock.

   Ecosystems services provide the ideal conditions for human beings and their societies to thrive; they maintain the life support systems for people. Without them life may not be possible. Without stable climate or fresh water, without crops forming seeds, how could we live?

3. Examples include: percentage habitat loss; year of maximum catch; classification of ecosystems services; increases in nitrogen flows; species extinction rates; freshwater extraction.

   The main value in the assessment is that it provides evidence for a need to move towards greater sustainability, which may allow politicians and environmental managers to make more informed decisions on priorities.

4. Whilst EIAs provide decision-makers with useful background information, the standards with which they are carried out vary considerably. In many countries EIAs are not required at all or regulations on how they are carried out allow developers to find loop holes. Some impacts may fall outside of the scope of the EIA, with some indirect impacts harder to assess. Whatever the findings of an EIA, socio-economic factors are likely to impact on final decisions made as the EIA process may add considerable expense to development.

5. They are useful tools to indicate sustainability, as they indicate two key elements: how much land we need for our resource supply and how much we need to absorb our wastes. When this total requirement for a society exceeds the land available to that society, then this indicates a lack of sustainability.

6. Pearce et al.: this quote is fundamentally focused on the economic aspects of sustainability, though this economic development must be within the boundaries set by ecological limits in order to be truly sustainable, so this is implied. Cultural and social aspects are not mentioned directly.

   National Academy of Science and Royal Society of London: this approach seems to focus on the ecological and the economic aspects, through considering the impacts of resource use. There is little mention of the social and cultural aspects.

   Forum for the Future: this statement has a social and culture focus, which presents human quality of life within the context of economic and ecological sustainability.

7. Any of these perspectives could be argued as better, but this will really depend on an individual's environmental values system. For example, the Forum for the Future could be argued as an improvement as it stresses that the actions of people are at the centre of sustainable development. This anthropocentric type of approach, about people and for people, understands that sustainable development must be related to and driven by human needs and quality of life.

8.

For	Against
Research has linked DDT to premature births, low birth weight, and abnormal mental development of infants.	WHO states DDT is safe if used properly.
Alternative methods of pest control exist; DDT is not the only available pesticide.	Alternatives are not as effective; DDT significantly reduces deaths from malaria when used.
Spraying cannot eradicate the 20 species of *Anopheles* mosquitoes from native habitats without disastrous impacts on other invertebrates and biodiversity.	Annual deaths from malaria are still over 1 million; 85% are in Africa. The number of malarial cases globally is over 240 million.

For	Against
The ecological effects are well documented, but the effects of accumulation in human tissue are not fully known.	Previous decisions to ban DDT saw a resurgence of mosquitoes and a rise in deaths from malaria in many countries.

9.

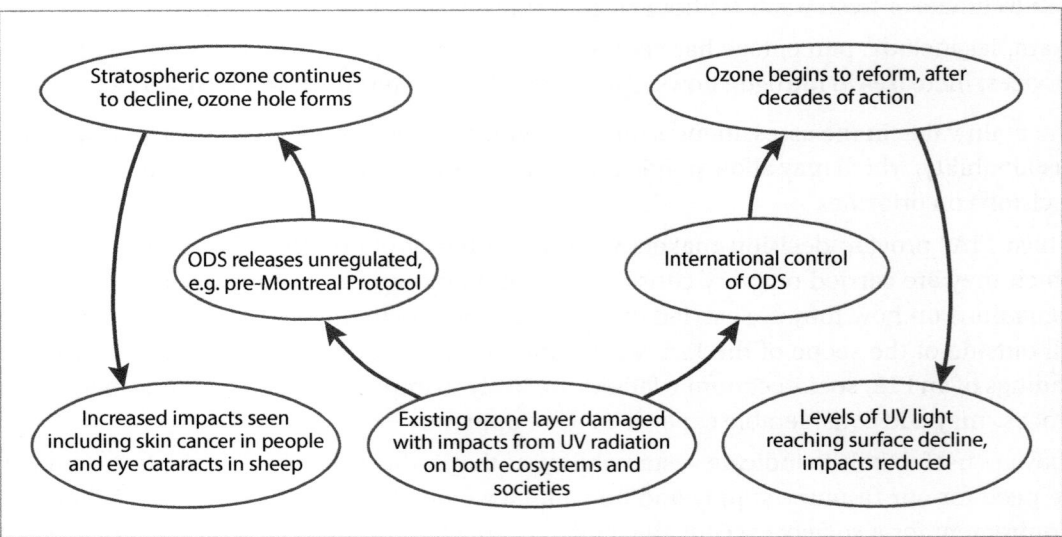

10. For example:

Cause	Agricultural	WHO global ban: promote alternatives
	Disease control	Promote alternative to DDT, e.g. malathion
Release/transfer	Agricultural	Aerial spraying good target for control.
	Disease control	Domestic use can be easily controlled, alternatives
Effects	Agricultural	Monitor for bioaccumulation effects
	Disease control	Research needed into bioaccumulation in humans

11. For example, the global ban in agriculture was largely successful, but alternatives are not always viable for malarial control. (For more, refer to the table in question 8.)

Chapter 2

Species and Populations

1. Ecology is the science of the relationships between living things and the non-living environment.
2. Ecosystems are complex systems that involve the interaction between the living and the non-living components of a defined unit. Ecosystems vary in scale, from the whole world, or ecosphere, to smaller systems like soil systems.
3.

Term	Definition	Example
Population	A group of organisms of the same species living in the same area at the same time and which are capable of interbreeding.	The number of Indochinese tigers, *Panthera tigris corbetti*, resident in South East Asia has been estimated at around 1,000 individuals.

ANSWERS TO REVISION QUESTIONS

Term	Definition	Example
Species	A group of organisms that interbreed and produce fertile offspring.	The tiger, *Panthera tigris*, is considered a single species with six subspecies surviving and three extinct.
Habitat	The environment in which a species normally lives.	The habitat of the Indochinese tiger, *Panthera tigris corbetti*, includes a variety of types of forest.
Niche	A species' share of habitat and resources in it. An organism's niche depends on where it lives and what it does.	The tiger's niche is that of a top carnivore, a major predator on animals like the wild pig, *Sus scrofa*, or other large mammal prey between 20–100 kg.

4.

Term	Definition	Example
Competition	Populations use resources such as space, light, mates, food, or nutrients that are finite. If they are used, there is less available to others and they may become limiting factors.	Duckweed and the mosquito fern compete for space on the surface of freshwater. In the UK, the otter and the mink are in competition for food around rivers.
Parasitism	This occurs when species live closely together, but one of the species gains at the other's expense.	Ectoparasites include leeches and mosquitoes. Endoparasites occur in many ecosystems. In the tropical reef environment, each species of sea cucumber is likely to be supporting a pearl fish.
Mutualism	These are relationships between species where both benefit.	Lichens operate as a mini-ecosystem. Coral and coral animals: the animals filter feed for floating detritus and the algae carry out photosynthesis.
Herbivory	This is the consumption of autotrophs by a primary consumer.	Limpets graze on seaweeds. Red leaf monkeys eat the fruit and leaves of fig trees.
Predation	This is the consumption of a primary consumer by a secondary consumer or higher.	Tigers and wild pigs live in rainforest.

5. A species' share of habitat and resources in it. An organism's niche depends on where it lives and what it does.

6. Advantages of model: It helps to graphically demonstrate a number of interactions between biotic and abiotic factors and how they may work together synergistically to influence the niche of an organism. This also helps gain understanding of the spatial relationships in the environment as the conditions can be mapped to distribution patterns. It is better than simpler models as it can indicate how more factors may work together giving mutual benefit or stressing the organism. Disadvantages of model: The model is a simplified representation of reality; it may therefore give a false sense of understanding. The real world is far more complex and other models that can show the interactions of more factors may be more useful for analysis.

7. At very small population levels of hosts, under HT parasites may be unable to find new hosts to colonise and may die out. At very small numbers there may not enough resource base to sustain a parasite population.

8. (a) From one full peak to another of either predator or prey.

ENVIRONMENTAL SYSTEMS AND SOCIETIES SL

(b) The predators reduce prey, and then the predator population falls due to lack of food. The prey population grows due to the lack of predator increase in intraspecific competition. The predator population then grows, and the prey populations falls back to the start.

Communities and Ecosystems 1

1.

Term	Definition	Example
Ecosystem	A group of populations living and interacting with each other in a common habitat.	In the community of a South East Asian rainforest, populations of tigers interact with other species through predation and competition.
Community	A community of interdependent organisms and the physical environment they inhabit.	Rainforest ecosystems

2.

Photosynthesis		Respiration	
Inputs	Outputs	Inputs	Outputs
Energy: light	Energy: chemical and heat loss	Energy: chemical	Energy: chemical and heat
Matter: carbon monoxide and water	Matter: glucose and oxygen	Matter: glucose or other carbon compounds	Matter: carbon monoxide and water

3. Refer to Figure 2.10 and Figure 2.11 in the main text.

4. Producers, in most ecosystems, are green plants or algae, photosynthetic organisms that form the base of the food chain. Consumers are organism that eats or gains nutrition from other organisms. Decomposers are fungi and bacteria that break food down outside their bodies, by secreting enzymes into the environment. They are important in recycling nutrients.

5.

1st trophic level	Autotroph	Primary producer	
2nd tropic level	Heterotroph	Primary consumer	Herbivore
3rd trophic level		Secondary consumer	1st order carnivore
4th trophic level		Tertiary consumer	2nd order carnivore
5th trophic level			Top carnivore
Decomposer subsystem		Detritivore	
		Decomposer	

Communities and Ecosystems 2

1. (a) people; (b) zooplankton; (c) fig/dipterocarp or bird's nest fern; (d) polar bear; (e) wild boar/leaf insect/leaf monkey/mouse deer; (f) krill/cod/seal/bear; (g) ringed seals.

2. See sections 2.2.11 and 2.2.14 for examples.

3. The first law of thermodynamics explains that there is 'no such thing as a free lunch'—energy cannot be created or destroyed. All food chains take in energy from the sun mostly by plant photosynthesis and process this through a series of energy transformations and transfers. The second law explains limited length to the food chain, as energy moves through a food chain there is always a loss with the amount depending on the efficiency. So, this limits food chain length.

Therefore, there is less energy at the end of a food chain and so larger predators at the top of food chains are always rarer. Therefore, there is less energy found in tertiary consumers than in the producers.

4. Due to the second law of thermodynamics, because energy cannot transfer with 100% efficiency, it must decline along a food chain. Food chains longer than six are rarer and indicate higher efficiencies of transfer from one trophic level to the next, often found in aquatic ecosystems (note species here cold blooded and supported by water, both of which are more energy efficient than being warm blooded and walking on land).

5.

Pyramid type	Standing crop	Productivity
Numbers	Total number	Numbers per hour, etc.
Biomass	Total biomass in g (kilo, micro etc.)	g/hour, etc.
Energy	Joules (kilo, mega, etc.)	J/hour, etc.

6. A pyramid of numbers, because there is great variation in size of organism it is easy to invert by having small herbivores or carnivores on larger producers or herbivores respectively. For example, caterpillars on a tree or mosquitoes on a zebra.

7. A pyramid of productivity, because irrespective of the relative size of an organism the pyramid of productivity show flow of energy over time. This can never be inverted, without breaking the laws of thermodynamics.

8. Data for pyramid of numbers, pyramid right.

	D	H	C1	C2
Numbers	684	571	138	3

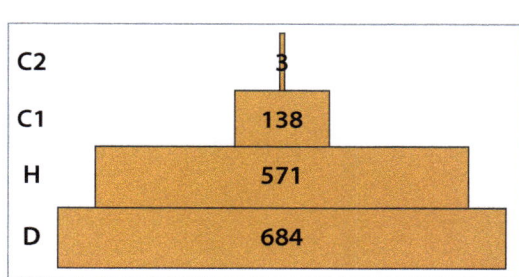

9. Example of calculations: Flatworms weigh 0.4 mg, approximately 5 are detritivores.

So, 0.4 x 5 = 2 mg to detritivore mass, true worms 0.7 x 37 = 25.9, add to total for detritivores, and so on.

10. Data for pyramid of biomass, pyramid right.

	D	H	C1	C2
Biomass	234	241	61	17

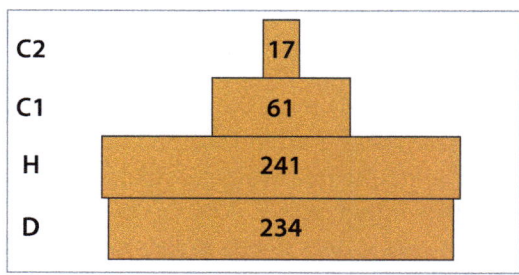

11. Make sure you effectively compare both pyramids using comparative statements, then explain the pattern you have identified. Here are some possibilities. Compare: the pyramid of numbers shows more detritivores than herbivores, whereas the pyramid of biomass shows more herbivores than detritivores. Explain: this indicates the detritivores are smaller in size than the herbivores. Compare: the pyramid of numbers shows a small proportion of C2 to C1 carnivores (3:128),

compared with the pyramid of biomass (17:61). Explain: the C2 carnivores are larger animals than the C1 carnivores.

12. The appearance of mouthparts such as biting, piercing or chewing, observed feeding behaviour in habitat or captivity, and gut or faeces content from dissection.

13. As for the earlier flatworm example, the number is divided equally between levels. This is a method limitation, which must be rarely true. In the above example most of the flatworms or even all could be detritivores. But identification to species level with invertebrates to establish trophic level is very time consuming and requires training.

14.

	Prey	Predator	% Eco Efficiency
Numbers	1,255	141	11
Biomass	474	78	16

Flows of Energy and Matter

1. Energy flows through the ecosystem; when it is used, it is converted to a new form. Energy cannot be recycled. Matter cycles through the ecosystem. The amount of matter is fixed (finite) and is constantly broken down and built up into new forms.

2. Because energy cannot be reused, it cannot be recycled. It is simply transformed from one type to another in accordance with the first law. According to the second law, in each transformation heat is lost which cannot be harvested by living organisms to reform chemical energy and exits the system, ultimately to space. Matter, on the other hand, is moved or transformed by energy, but it doesn't follow the same rules. It can be reused and it can be transferred with complete efficiency.

3.

$MJ/m^2/year$	Herbivores	Carnivores
Spitsbergen	12.5	1.25
London	16	1.6
Miami	32.5	3.25

4. There are many losses as solar energy enters the earth ecosystem including reflection and transmission by leaves. Some light may not land on leaves, absorbed or reflected by clouds, hitting bare ground etc. All light is not available to use for photosynthesis – just restricted wavelengths. For example, all green light is reflected by a green leaf, and this energy cannot be used. Some light energy is absorbed into the leaf and transformed to heat energy and therefore not useful for photosynthesis. Photosynthesis itself is also enzyme limited, and works at a slower rate if temperatures are low.

5. The polar bear needs more tundra as the low insolation means that the Arctic ecosystem has a low GPP therefore a limited resource base at the start of the polar bear food chain. Conversely the tropical ecosystems have a much higher rate of photosynthesis and an all year round growing season. GPP is much higher. Polar bears (around 200–400 kg) are much larger than sun bears (45–60 kg), so one polar bear needs more energy than one sun bear to stay alive.

6.

	Full Name	Definition
GP	Gross production	This is equivalent to food assimilated, and so it is equal to the food ingested minus faeces lost. It is what occurs before respiration.
NP	Net productivity	This is the growth of an organism. It is what occurs after respiration.
R	Respiration	The process in which useful energy is released from food.
NPP	Net Primary Productivity	The actual growth of the producer, the part of GPP that is available after the producer has carried out respiration.
GPP	Gross Primary Productivity	The amount of photosynthesis carried out by the producers.
GSP	Gross Secondary Productivity	The amount of food eaten minus the faecal loss.
NSP	Net Secondary Productivity	The actual change in biomass.

7. NPP (average) = 962 mg/m³/hr

   R = 2,245 mg/m³/hr

   GPP = 3,207 mg/m³/hr

8. (a) NSP = (390.8 - 380.6) = 10.2/8 = 1.3 g/day

   (b) Food and faeces = 10.3 g. Assimilated food = 268.9 − 10.3 = 258.6/8 = 32.3 g/day

   (c) R = 32.3 − 1.3 = 31.0 g/day

9. As matter is finite, it has to be the same material on earth. Energy cannot stay, due to the second law or thermodynamics heat is produced which radiates out to space and is permanently lost from the system. So dinosaur matter may be present in your body, but not dinosaur energy which has been lost.

10. To get to the rabbit lung tissue the nitrogen must travel through the food chain. Note: nitrogen is in large amount in the atmosphere, but there are limited pathways to bring it into the ecosystem through nitrogen fixing bacteria, lightning (Haber process) and fertiliser production. Nitrogen cannot be directly absorbed by animals.

11. The energy source is solar/sunlight. Transformation processes are photosynthesis, respiration, combustion, decomposition. Transfer processes are consumption along food chain, detritus movements.

Biomes, Zonation, and Succession

1. The amount of light energy per unit area varies, in relation to latitude; more light energy arrives at lower latitudes throughout the year.

   This occurs as the same amount of solar insolation is spread over a wider area at high latitudes but concentrated in a smaller area at the equator.

   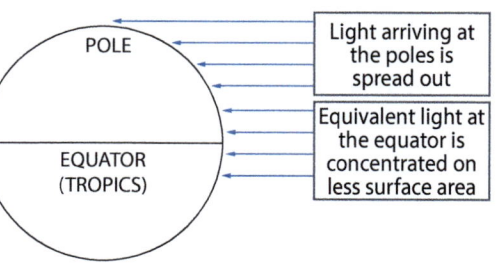

2. Plant photosynthesis, and therefore growth, is lower where less insolation energy is available and vice versa.

3. The character of biomes is largely dictated by the vegetation type found there. This vegetation type is also a response to the climatic conditions for production by photosynthesis. Temperature influences the rate of photosynthesis.

4. **Tundra:** There are no trees as the soil water is permanently frozen—a permafrost. Growing rates are very low overall. **Deciduous forests:** Trees grow well, although not as tall as in the tropics. The biome found here is because of rainfall and seasonal temperature variations. **Desert:** Found where dry falling air naturally occurs. The desert belt is found at around 20–30 degrees north or south of the equator. **Tropical Rainforests:** A tall, thick canopy at around 40 m, with emergent trees up to 80m. It grows where it rains every day, with no dry season. Rainfall is greater than transpiration.

5. Glacier Bay, Alaska: pioneers such as Dryas colonise quickly, followed by slower growing trees species like willow and eventually slowest growing species in the spruce mixed forests at the end. From r-type to K-type.

6. Positive feedbacks occur at each transition to the next seral stage. The pioneer community produces deeper soils allowing a willow scrub to establish, which in turn shades out the pioneers preventing them from growing. So deeper soils produce deeper soils, allowing bigger and bigger plants to grow as capacity for primary production increases and forest develops.

7. Strengths of the method include that the technique allows for collection of meaningful data to describe the patterns, without being overly time consuming. The weaknesses of the method will depend on how exactly it is applied, but abiotic factors change through daily cycles, seasonal cycles, short term weather patterns, and tidal changes.

8. (a) It starts low, rises, then falls.

    (b) There is a loss of diversity along most of the line, particularly seaward side.

    (c) There is a lack of competition, and the species is effective at colonising sand.

    (d) Succession, over time. Succession stages (seres) mimic the zones.

9. A baseline survey should provide a description of an ecosystem that can be referred back to later to see what changes have occurred. A baseline survey could be carried out by using a number of interrupted belt transects.

Investigating Ecosystems 1

1. Taking the swamp forest of the Bangkok delta flood plain around Phra Samut Chedi as an example, the abiotic factors that influence the vegetation are light, temperature, salinity, water depth, substrate type, flow rate, wave action, pH.

2. With reference to the same example, the three most significant factors affecting the vegetation are salinity, water depth, and substrate. Where water is fresher, water species such as the nipa palm (*Nypa fruiticans*) dominate, and higher salt concentrations lead to mangrove species such as white mangrove (*Avicennia*) and red mangrove (*Rhizophora*) having an advantage. In order to measure how these factors alter how transects should be used, you have to start at the top of the zonation pattern and sample at regular intervals from the land. Intervals should be taken at regular intervals of depth, not distance from the land. The depth measurements can then be correlated with distance, salinity and substrate type to determine from a number of different transects what the key factors are in different locations.

3. When evaluating the sampling programme try to go beyond 'take more samples'. Evaluation should consider the location and numbers of samples in respect to the habitat. How varied are they? Are they representative of the habitat? Are the results consistent?

4. Key example to some British temperate deciduous trees

    (a) Is the leaf simple?

    >   Yes: go to 4

    >   No: go to 2

(b) Is the leaf palmate?

   Yes: (H) Horse Chestnut

   No: go to 3

(c) Are the leaflets slightly or deeply toothed or serrated?

   Slightly serrated: (B) Ash

   Deeply Serrated: (G) Rowan

(d) Is the leaf cordate with a drip tip?

   Yes: (D) Small leaved lime

   No: go to 5

(e) Does the leaf have deep notches or lobes or is it smooth or toothed?

   Deep notches or lobes: go to 6

   Smooth border or toothed: go to 7

(f) Does it have deep angular notches in a simple, palmate leaf? Yes: (C) Sycamore

   Is the leaf simple with deep round notches? Yes: (A) Oak

(g) Are the borders of the leaf smooth or toothed?

   Simple and smooth: (E) Cherry

   Simple and toothed: (F) Hornbeam

5. (a) 25%
   (b) 68.9 kg
   (c) 423 g/m²/year
   (d) 1.16 g/m²/day
   (e) 5 ha = 50,000 m² × 423 g/m²/year /1000 = 21,150 kg
   (f) Strengths of method: It gives a good indication of the rate of growth, which can be useful for constructing pyramids of productivity or trying to estimate sustainable yields. It can also be applied to agriculture. Weaknesses: results are likely to be specific to this site. To get a better understanding of grasslands like this, many sites are needed for replication to understand general patterns of productivity in grasslands. An increase of sample size from 25 quadrats would improve the reliability of the data. As results were repeated elsewhere then the soil conditions and grass species could be different, giving varied results.

6. It will vary from 20–40%, depending on the individual. Estimates are subjective measures.

7. 69/100 = 69%. Note: this is not a useful measure as the sample unit is too large, consider when evaluating.

8. 20/121 = 16.5%. Note: this time the frequency measure becomes a useful indicator of abundance, because the quadrat sample unit is smaller.

9. Estimates of percentage cover are quick and easy to take, but they can be very subjective. Therefore, it may not be useful to compare work from different people. Training and experience can give more consistency. Measurements of frequency are objective indicators of abundance, but can be time consuming.

Investigating Ecosystems 2

1. (a) M = 131, n = 143, r = 27, N = (131 × 143)/27 = 694

   (b) For this example, confidence limits are high as r is over 18% (when r is very low confidence is low).

2. (a) A
   (b) A = 7, B = 2.8
   (c) B

Chapter 3

Biodiversity

1.

Term	Definition
Biodiversity	The amount of biological diversity or living diversity per unit area, it includes the concepts of species diversity, habitat diversity, and genetic diversity.
Species diversity	The variety of species per unit area. This includes both the number of species and their relative abundance in the community.
Habitat diversity	The range of different habitats or number of ecological niches per unit area in an ecosystem, community, or biome.
Genetic diversity	The range of genetic material present in a gene pool or population of a species.

2. There are many difficulties in achieving an accurate list of all species, as well as difficulty in accessing some areas, there can be difficulty in identifying many taxonomic groups and a shortage of specialists.

3. (a) Site 4 must have one or two species dominating the diversity, whereas the 13 species in site 5 must be evenly distributed.

   (b) Some sites are not clearly related, for example, site 4 has low diversity but high RHI, whereas site 2 has higher diversity but a lower RHI.

   (c) (i) As tourists are present then RHI type approaches could be carried out by tourist divers. Physical damage will be shown well by this method.

   (ii) Coral bleaching may be better monitored by specialists, though the damage can also be conspicuous as it is a colour change that doesn't need species identification. However, which species are impacted is important.

   (iii) This has to be carried out by experts, otherwise the data would be meaningless.

Biodiversity and Conservation

1. Natural selection is one of the main mechanisms by which evolution occurs through the following steps:
   - Within a population of one species there is genetic diversity, which is called variation.
   - Due to natural variation some individuals will be fitter than others.
   - Fitter individuals have an advantage and will reproduce more successfully.
   - The offspring of fitter individuals may inherit the genes that give the advantage.
   - Over time, the species changes in the direction of selection, becoming fitter. With many generations and a long period of time speciation can occur.

2. See section 3.2.3. For example, research detail of evolution in Darwin's Finches which became isolated on different islands and adapted to the conditions. For a top-level answer, see recent research on Medium ground finch, *Geospiza fortis*, found on the Galapagos island of Daphne Major.

ANSWERS TO REVISION QUESTIONS

3. See Table 3.2 and the summary text in section 3.2.5.

4. See Table 3.2 and the summary text in section 3.2.5.

5. (a) There is an increase in diversity; at climax, in many ecosystems, there is a slight decrease.

   (b) This is a principle applied in nature reserve management. A greater variety in habitat gives a greater variety of niches, and all else being equal, greater diversity.

   (c) This is shown for example in complex ecosystems such as tropical rainforests. Rainforest ecosystems appear to be stable systems in terms of inertia, e.g. resistant to invasion by foreign species.

   (d) Human activities such as burning, grazing, and logging may modify patterns in succession leading to a plagioclimax. Examples of plagioclimax include heather moorlands in the UK that are burned and many areas of grazing land which were originally forest. In many instances, these human activities often simplify ecosystems, which could make them unstable, e.g. North America wheat farming versus the native tall grass prairie found in the Great Plains

6. Meteorite impact, volcanic eruptions, droughts, ice ages, climate change and sea level rise, and genetic inferiority.

7. Extinction rates are calculated from an incomplete fossil record, which may use information predicted from families or genera rather than the species itself. As we don't currently know the total number of species on the planet, it is impossible to be certain about the rate of extinction of species. Current day predictions are based on what has been observed, but some surviving species may exist in remote locations or where expertise to identify the species doesn't exist.

8. The current rate of extinction is at least 100 times the background rate and possibly as high as 10,000 times the background rate. It is generally accepted that we have entered the sixth mass extinction which differs from the previous five events in that it has been caused by humanity.

9. A HIPPO

    - Agriculture
    - Habitat destruction and degradation
    - Invasive non-native species
    - Pollution
    - Population
    - Overharvesting

10. Ultimately habitat destruction and degradation are the most significant as, if large areas of habitat are intact, all the other human impacts are less likely to cause extinction. Note: it is possible to argue others in specific contexts, e.g. in a degraded and fragmented habitat hunting may be more significant, or in island communities then invasive species have caused large numbers of extinctions.

11. Tropical forest biomes are seen as particularly vulnerable ecosystems, both due to deforestation rates and slow regeneration. Rainforests are valued for their high biodiversity and are home to over 50% of the planet's species, many of which are endemic. Tropical marine biomes also contain high biodiversity with can be under threat. Coral reefs are the most diverse of marine ecosystems, but are vulnerable due to human impacts such as overfishing, mangrove removal, climatic change causing bleaching, and the impacts of ocean acidification. See section 3.3.8.

12.
- Population size and reduction over ten years
- Geographic range
- Area of occupancy
- Number of fragments
- Numbers of mature individuals
- Probability of extinction

Use the first column of table 3.5 to outline a given factor. For example, if the population size has reduced by over 90% in ten years, the species would be critically endangered.

13. Refer to Table 3.6 on page 91.

14. As an example, the biodiversity of the evergreen forests of Northern Thailand are under threat from commercial deforestation, an increasing human population in the area, human fires, and climate change. These have resulted in many extinctions. The forests also have an important hydrological function as watershed forests help to maintain dry season stream flow. But, due to deforestation and increased extraction from the steams, increasingly, the dry season flow is failing; this may have devastating consequences for mammals that rely on this water to survive.

15. Refer to Table 3.8 on page 94.

16. Refer to Table 3.9 on page 95.

17. They are widely seen to have failed by conservationists, as stated by the meeting of the CBE at Nagoya in 2010, which concluded that the targets had not been met and the loss of biodiversity continues. However, this is not to state that any of the agreements have had no positive impacts. The positive aspects would include that all the strategies have raised awareness of the issues and situation, increased the number of protected areas and help for individual species. Nagoya concluded that the fair sharing of benefits had not been achieved, but it consequently remains a current focus.

18. Use the principles of island biogeography outlined in as applications in Table 3.10 on page 97, e.g. larger reserves are better than small ones, as the rates of extirpation (local extinction) or species depends on habitat size.

19. Mink are opportunist predators, meaning they take what they can. In the presence of otter, they change their prey items to take fewer fish and more birds. In the absence of otter, it is likely they would take more fish in their diet and fewer birds and mammals.

20. This can be argued either way.

    For: Mink should be culled as they are not a natural part of the native fauna at Slapton Ley, and they are predating on the bird and small mammal community, which the reserve is there to conserve, and some of which may be rare enough to be threatened by the predation. Impacts on the otter are not seen as significant, but there may be some influence that the studies have not shown clearly yet.

    Against: Mink should not be culled as they are filling an empty niche in the community and represent a useful part of the ecosystem. They are an attractive and interesting member of the reserve fauna, which increases the aesthetic value of the reserve. Mink have an intrinsic right to exist, irrespective of if they are 'natural' in this habitat; any culling that intends to remove the mink population is against this intrinsic right to exist and the individual animal rights.

21. For example, the Slapon Ley NNR area has had considerable success in protecting an extensive habitat of reed beds, marshland, wet meadow, and open water. It also includes a significant shingle ride habitat. A number of significant species have been protected and thrived on the reserve, including strapwort, the Cetti's warbler, and otter. Criticisms of the success of the reserve are that it focuses on a mixture of traditional and scientific approaches to active management, but a wilderness approach may have been equally successful and could lead to a more interesting

ANSWERS TO REVISION QUESTIONS

conservation area, in which previously extinct species like the European beaver could be re-released. Water quality shows continuous decline and eutrophication remains a threat. There is an inability for reserve managers to regulate farming activity in the upstream catchment areas. The reserve includes a road that is being protected from rising sea levels; arguably, it is more natural to follow a managed retreat leading to a the loss of the freshwater lake and development of salt marsh habitats.

22. See section 3.4.5.

Chapter 4

Water and Aquatic Food Production Systems and Societies

1. Water on the planet is mostly stored in the oceans, with only a small proportion as freshwater. Most of the freshwater on the planet is in glaciers and ice caps. On land, there is great variation of water supply due to climatic factors. Coastal regions nearer to warm ocean currents have greater precipitation as wind blows in from the sea.

2.

Term	Definition
Atmospheric store	Water vapour in atmosphere supplied by ET from surface.
Condensation into cloud store	Water vapour condenses to liquid water in clouds due to temperature change and the presence of dust or other particles (nucleating surface).
Precipitation to surface	Precipitation falls due to the build-up of water droplets, their combining to greater mass, and then falling as rain or snow.
Water storage	Stores include the bodies of all living organisms, or non-living storage such as atmosphere, ice cap, ground water, lakes, and ocean.
Rivers and glaciers	Flows of water overland as ice or liquid.
Evapotranspiration (ET)	ET is the combination of evaporation and transpiration; it requires both heat from sunlight and the biological action of plants.
Convection	Water carried by hot, moist rising air higher up in the atmosphere.
Advection	Wind-driven movement of water horizontally through the atmosphere.
Sublimation	Snow and ice moving directly to atmospheric storage from solid to gas.
Interception	Precipitation may be intercepted by vegetation or hit the ground.
Surface water	Water storage on the ground surface.
Run-off	Movement of water over the ground surface to rivers.
Infiltration	Movement of water from surface through the soil.
Percolation	Movement of water through porous rock and sediment.
Throughflow	Movement of water through soils to rivers.
Groundwater	Storage of water in rocks underground (aquifers).

3. See Figure 4.1 on page 106.
4. Examples are given in Table 4.4 on page 109 and strategies in section 4.2.5. For Ogallala aquifer, solutions would be to change farming practices, so it demands less water. For the Aral Sea, a solution would be to improve irrigation, as it may help to reduce extraction rates. Tuvalu rainwater

harvesting is a solution and a problem, as it prevents aquifer recharge. Desalination may help in this situation.

5. A case study of an international water source conflict can be found in section 4.2.8.

6. See Table 4.6 on page 115 for an example of the history of the Grand Banks cod fisheries off the coast of Newfoundland. A solution to improve the sustainability of the fishery would be to apply the theory of maximum sustainable yield.

7. Aquaculture supplies over 50% of fish for human consumption: without aquaculture human consumption of fish would have been unable to rise without causing marine species to become extinct. On the other hand, aquaculture can in some cases increase pressure on ocean fisheries where large inputs of wild fish are used for feed, and it can also have damaging ecological effects. For a case study, see section 4.3.10.

Water Pollution

1. BOD stands for biochemical oxygen demand. It is a measure of the total demand for oxygen by living and chemical components in a water body.

2. It is normally measured over a five-day period. Initial oxygen concentration is taken from a water body and samples are collected in sealed bottles with no air spaces. The samples are kept in a dark cupboard for five days. The oxygen content is then measured again, and the difference is calculated in mg/l.

3. (a) The second site is more polluted by organic waste from agricultural and domestic sources. In the second site, lower oxygen levels mean oxygen demanding species can't survive at this site. Species more tolerant of low oxygen will be greater in abundance. This combination results in a lower biotic index (index of 4.6) for the second site.

   (b) The second sample has a lower diversity according to the Simpson's index. There are fewer species in total, and the community of the second site is dominated by swimming mayfly. This diversity index comparison also indicates that the site is polluted.

   (c) Biotic indexes provide an indirect measure of sewage or other organic pollution. Species indicate environmental conditions due to their tolerance limits across their whole life, this can be better than direct as it indicates the water quality over longer time spans. However, many variables are involved in species distributions, direct measures are more robust and certain. Using both together makes sense.

4. Eutrophication is the process of nutrient enrichment of an ecosystem. See Figure 4.6 for the flow chart (a simplified version is sufficient to demonstrate why oxygen falls).

5. Positive feedbacks:

   (a) Increase in nutrients within system, increase in primary and secondary productivity, increase in DOM, increase in nutrients within the system.

   (b) Increase in DOM, increase in decomposers, increase in BOD, decrease in oxygen, increase in death rates of gill breathing fish.

   (c) Increase in turbidity, increase in invertebrate death rate, increase in DOM, increase in nutrients cycled in system, increase in planktonic algae, increase in turbidity.

6. The impacts of eutrophication include reductions in biodiversity, water quality, long-term negative impacts on fisheries (though, in the short term, this may include a reduction in the recreational value of a lake and environmental quality in general). The most significant impacts, and most immediately noticeable to society, are often first aesthetic, such as fish death incidents in large numbers, such as Lake Erie in the 1960s. Longer term issues include the development of dead zones found, for example, in the Gulf of Mexico.

7. See Table 4.12 on page 124.

8. Address the following points:
   - Difficulties of influencing human behaviour without legislation or economic incentive
   - Catchment management agreements may be difficult to monitor
   - Changes may be costly
   - Disturbing habitats may have negative effects on biodiversity
   - Consequences of biomanipulation may not be predictable
   - Point source pollution control is easier to apply and monitor
   - Phosphate is the limiting nutrient for nitrogen-fixing blue green algae
   - Phosphate mainly transfers through detergent, sewage and surface run-off as it is non-soluble
   - Nitrate dissolves readily and can follow all pathways in the hydrological cycle.
   - Pumping mud may have other impact on the water quality (for example, increasing turbidity)
   - Barley straw bales inhibit blue-green algae growth but do not prevent other algae

Chapter 5

Soil Systems and Terrestrial Food Production Systems and Societies

1. Soil systems are at the boundary between the living community and the abiotic environment. They are essentially ecosystems themselves. The biome or ecosystem above influences the type of soil directly by influencing the rate of biological weathering and nutrient cycling. Decomposers and detritivores in the soil community help to break down organic matter and produce acids, such as humic acid, that further weather the bedrock.
2. B would be approximately 53% clay, 24% sand, and 23% silt.
3. See Table 5.1 on page 130.
4. See Table 5.7 on page 136.
5. See the case studies of 5.3.4 and 5.3.5 for examples.
6. Food overproduction and waste is associated with more developed countries, rather than developing countries. Incidence of major famine in the last 50 years has all been in the developing countries, particularly countries in Africa and Asia. Diseases of malnutrition also show a clear and uneven global distribution pattern. The populations in MEDCs suffer greatly from obesity whereas the poorest countries suffer most from protein-energy malnutrition.
7. The quote suggests that, in the long run, our food supply is not sustainable if it is to rely on oil. Improvements in agricultural productivity have been dependent on fossil fuels to produce fertiliser, for machinery and irrigation. When global peak oil production is reached, it is likely to cause a rise in prices that will mean food will become more expensive, or productivity may decline unless alternatives are found to fossil fuel support of agriculture.
8. In terrestrial systems food is usually harvested at the level of producer, or herbivore. These crops are more efficient in terms of the trophic level occupied by the food source. In aquatic systems carnivorous fish are often harvested, such as tuna or salmon. As they are further up the food chain (they are high order carnivores), there is a great reduction in energy stored in their trophic level. Aquatic systems are often more efficient in terms of energy transfer; however, the amount of sunlight entering the base of the food chain in the producers can also be reduced because sunlight is absorbed with depth in sea water. Much of the ocean surface is also limited as nutrient availability limits productivity; nutrients (particularly iron) sink down into the deep ocean. Each case of terrestrial versus aquatic systems yields different comparisons. Mariculture systems that are used to cultivate oysters or algae (seaweeds) for consumption are more efficient than open ocean fishing for carnivorous fish. Farmed stock can be compared in terms of the food conversion

ratio, a measure of the animal's efficiency in converting feed to body mass. For example, 2.2 kg of food is used to produce 1 kg of chicken, compared with 12.7 kg for 1 kg of beef. Whereas many aquatic systems have a greater efficiency, with prawn farming ranging from 1.2–1.65 kg of food for a kg of live weight.

9. Here Asian subsistence agriculture is compared with US cereal farming. These two systems are different in terms of the degree of openness, in other words, how much the system requires inputs and outputs from far away. Subsistence farming uses the available resources supplied from the local area as input, as output the food is used to supply local needs. The commercial system involves more import of resources from further away, with machinery, fertiliser and labour brought in to the area. Outputs supply a national or global demand through distribution systems operated by supermarkets and using long distance haulage.

10. Use Table 5.3 on page 133 to compare inputs and outputs, and section 5.3.4 to review impacts of different strategies (Asian subsistence agriculture vs US Commercial or other contrasting case studies could be used).

11. Judged against the indicators of sustainable agriculture, subsistence farming appears to be the more sustainable. However, only small numbers of people can be sustained per unit area from this kind of agriculture given the limited total output. If this was more widely practiced, then greater areas of land would be needed to meet demand for food, and supply of food to urban areas would become difficult. The impact on destruction of habitats and conservation of biodiversity would be considerable.

12. Refer back to the discussions in sections 5.2.3 and 5.2.4.

Chapter 6

Structure and Composition of the Atmosphere / Stratospheric Ozone Depletion

1. Nitrogen, oxygen, argon, carbon dioxide, and water.
2. See Figure 6.1 on page 145.
3. Ozone is formed in a series of chemical reactions that occur when ultra violet light collides with oxygen. Ozone molecules are destroyed and reconstructed in the atmosphere continuously as UV light is absorbed. If the ozone layer is undisturbed by pollution (by halogens), it will continue to maintain a protective layer over the Earth which will remain in steady state or dynamic equilibrium, changing, with seasonal max and minimum, but maintained as a functioning ozone layer.
4. Halogens such as chlorine and bromine react with ozone (II) and monoatomic oxygen (II). E.g. chlorine: one single atom may cycle through these reactions 100,000 times, disrupting the normal equilibrium of ozone formation and destruction. The equilibrium for ozone levels shifts (you can show this with equations).
5. It increases mutation rates in DNA, causing skin cancers in humans. It can cause eye cataracts, damage the ability to carry out photosynthesis, reduces primary production, and also therefore can reduce total productivity.
6. See Table 6.5 on page 155.
7. Any from this list: replace ODS dependent refrigeration with existing 'Greenfreeze' technology driven by propane or butane; pump action sprays instead of aerosols; alternative propellants to ODSs; alternatives to aerosols, e.g. shaving soap instead of spray foam; alternatives to CFCs; alternative packaging; alternative pesticides; organic farming.
8. International organisations have been essential, as a global pollution problem ozone depletion required coordinated international action. This came through the collaboration of the scientific community and governments with the UNEP as the coordinating body. The chemical industry

complied with the regulations and developed alternatives. The Montreal Protocol has allowed for unprecedented economic and political cooperation and is seen as a success story by many environmentalists. International NGOs such as the EIA have helped to police illegal trade. Local groups have helped to raise awareness and promote action within countries. In the UK the national NGOs such as Friends of the Earth helped to raise awareness during the 1980s and Greenpeace campaigns in 1996 helped demonstrate that 'Greenfreeze' technologies are viable options for commercial usage.

Photochemical Smog

1. At ground level, ozone is formed as a secondary pollutant by reactions involving the nitrogen oxides (NOx). NOxs are formed from reactions of atmospheric oxygen and nitrogen in internal combustion engines, most commonly cars. In particular, nitrogen monoxide (NO) and nitrogen dioxide (NO_2) are formed. NO_2 reacts with sunlight forming ozone—hence 'photochemical'.

2. Ozone can cause lung damage, irritate the eyes, damage plants and corrode fabric. It is a powerful oxidant and reacts corrosively with many materials.

3. Smogs are products of weather combined with pollution. Photochemical smogs require sunny, windless conditions so that primary pollutants are not dispersed, and sunlight drives the reactions. Urban areas are often hotter due to lack of vegetation and water and the ability of concrete to absorb heat. This urban heat island effect may help to reduce dispersal and set up temperature inversions.

4. See Table 6.3 on page 151.

5. Overall it is better to control the pollutants at source by reducing the total number of emissions. Promoting clean air technologies in urban areas, for example, clean fuel burning, can help encourage change. Through monitoring and regulating industry and vehicle emissions, impacts can be reduced, though there may be unwanted side effects to technologies, such as catalytic converters in cars that make them less fuel efficient. Changing behaviour may help to reduce the impact of air pollution on human health, but ultimately this is not controlling the problem.

Acid Deposition

1. Acid precipitation refers to wet forms of acidic pollution (there is also dry acid deposition). The most common source is fossil fuel combustion that leads to sulphur dioxide (SO_2) and nitrogen oxides (NOx) forming. Sulphur dioxide and nitrogen oxides combine with any form of atmospheric water to form acid precipitation, most commonly rain containing sulphuric and nitric acids which can have a pH as low as 4: 'acid rain'.

2. See Table 6.4 on page 154.

3. In contrast with global pollution problems, such as ozone and climate change, or local ones such as urban air pollution, acid deposition problems are regional. Acidified deposition can travel easily across natural boundaries from one nation to another. Global problems have a global impact irrespective of who is doing the polluting; impacts of stratospheric ozone depletion were first noticed, in particular, over the Antarctic continent, though most of the pollutants were released in the Northern Hemisphere. Local problems involve pollutants that have not travelled far, which is particularly relevant in managing urban air quality.

4. See Table 6.5 on page 155.

5. Agreements can be reached between countries about the causes of acidification and, possibly, compensation. These may be bi-lateral, e.g. the 1991 Air Quality Agreement between the US and Canada focused on acid rain and was later expanded to cover smog. Some agreements may cover larger regions, e.g. the 1999 Gothenburg agreement to abate acidification, eutrophication, and ground-level ozone in Europe.

ENVIRONMENTAL SYSTEMS AND SOCIETIES SL

6. Overall, it is better to control the pollutants at source by reducing the total number of emissions; this may be achieved, for example, by encouraging energy conservation. Legislating for clean air technologies—for example, only burning low sulphur coal—can encourage change. Though monitoring and regulating emissions impacts can be reduced, there may be unwanted side effects to technologies such as scrubbers leading to mining of limestone and associated environmental impacts of this process. Treating impacts on ecosystems—for example by lime bombing—may have other effects and requires constant maintenance.

7. The level of cause is the ultimate way to reduce the impacts of pollutants: if less is generated, then this must be better. However, there are two ways to achieve this and they are not necessarily the same. The first is by a total reduction in demand, say for energy, as less pollution will be produced. The second is by changing the source of the pollutant, e.g. by banning CFCs, as then the problem is postponed rather than solved, as replacements have an impact on the ozone layer and may have other impacts, in this case as greenhouse gases. Other strategies of treating the pathway and the effects are really treating the symptoms following the event, they can be seen as useful tools in reducing the impacts, but rarely do they stop the problem. Consequently, whether it is regular lime bombing of a lake, or advising an urban population to stay indoors, they are not permanent solutions to the problem.

Chapter 7

Climate Change and Energy Production

1. Consider a range of options and applications for transport and electricity production. The main divisions are:

Oil, Coal, Natural Gas	All can be used for producing electricity and powering transport.	Commonly used as stored energy; they can be released in large amounts and quickly. Steam turbines are used to generate power.
HEP, Geothermal, Wind	All can be used to produce electricity in situ. Production depends on site specific details. Not all are usable for transport unless electric.	Turbines used to generate in many, e.g. geothermal and solar thermal use steam. HEP and wind use direct kinetic energy to turn the turbines.
Solar		Photovoltaic cells allow direct conversion of solar energy to electricity
Nuclear Fusion and Fission	Fission is available for electric power generation, but fusion is still in the development stage.	Fission uses turbines to generate electricity.

2. A range of comparisons are possible, for example:
   - The UK uses nuclear power; Thailand doesn't yet.
   - Thailand imports more electricity than the UK.
   - Thailand uses more alternatives than the UK.
   - The UK has a higher energy demand in total than Thailand.

3. The UK is a developed country with a nuclear power industry that arose in connection with research and development of nuclear weapons. The history of this choice can be linked to political drive to develop both the weapons and the power. A skilled and suitable trained work force is required to develop nuclear power. Thailand has not yet moved in that direction, though it remains a possibility as the country develops economically and improves its higher education. Thailand has a suitable landscape and climate for HEP, though seasonality can reduce power output in the dry season. In contrast with the UK, Thailand has a larger land area and uses many dams to help control seasonal flooding in the wet season.

ANSWERS TO REVISION QUESTIONS

4. Both countries have limited oil reserves, though they both have oil fields. Both countries also have associated gas fields that are used more than the oil reserves. Oil is expensive and is likely to have a great value for export and cheaper alternatives can be used for power generation. UK still exports oil to France, Germany and Netherlands, though, on balance, it is a net importer.

5. Remember: when answering compare questions, always be clear about what you are comparing. A table with comparative columns can help with this. The comparisons must make sense across rows. See examples on the next page.

HEP, e.g. Pak Mun	Coal, e.g. Bor Nok	Nuclear, e.g. Kalasin
Relatively small amount of electricity produced	Potential to produce a large amount of electricity	
Renewable resource, does not require import of fuel	Requires import of coal from Indonesia at additional cost	Nuclear fuel needs to be imported
Some greenhouse gases produced from decomposing in reservoir	Massive amounts of greenhouse gases, especially carbon dioxide	Very low carbon dioxide emissions
Relocation of houses and damage to fisheries	Damage to agriculture (pineapple growing) and squid fisheries.	Public opposition is strong, with demonstrations in proposed sites
Water pollution: acidity change	Water pollution: thermal pollution from cooling system	Possible radioactive contamination from waste and coolant water

6. Energy choices for Thailand. Availability: large supply of natural gas that is used for transport in Bangkok. Economic: invested money in a number of expensive HEP and Coal power schemes. Running costs on HEP have remained higher than predicted. Cultural change has produced increasing power demand, as air conditioning and hot water supply have become the social norm. Environmental considerations are large, as protest movements have found some success in stopping coal and gaining compensation for displacement for HEP. Technological: nuclear has been proposed, but Thailand does not have a trained work force to handle nuclear power stations yet.

Mitigation and Adaptation

1. The greenhouse gases allow short wavelengths of radiation, such as visible light and UV, to pass through to the Earth's surface, but they trap the longer wavelengths, such as infrared radiation, that are radiated from the earth. This is called the greenhouse effect and it regulates the Earth's temperature keeping conditions warmer than without the effect and more hospitable to life.

2. Carbon dioxide levels and temperatures have fluctuated over time. The atmospheric conditions in the earliest atmospheres of around 3,500 million years ago shows carbon dioxide levels as high as 7,000 parts per million, fluctuating in geological time down much lower to under 200 ppm. It should be remembered that comparisons with records from longer ago represent a world with very different solar inputs, climatic and oceanic circulation, and living organisms. This more recent and useful comparisons with today's climate are from the ice age periods, the temperature and carbon dioxide fluctuations of the last two million years. This period includes several long ice ages and warmer shorter interglacial periods. The maximum is around 300 ppm, and the minimum is around 180 ppm during the glacial stages. In comparison, today's concentrations are over 400 ppm which is higher than any of the other warm periods in the ice ages.

3. See Table 7.7 on page 166.

4. All of the greenhouse gas concentrations in the atmosphere have shown a dramatic increase in recent times, since the mid-20th century. The reasons are summarised in Table 7.7.

5. Historical records show warming over the last 100 years, from an average below 14°C for the first three decades of the 20th century, to an average over 14°C for the last three decades. The warming for the 20th century until the end of the century has been estimated at around 0.8°C.

6. Methods include: analysis of air trapped in ice from ice core data; dendrochronology from tree ring data; pollen analysis from sediment cores.

7. An example of a negative feedback flow diagram:

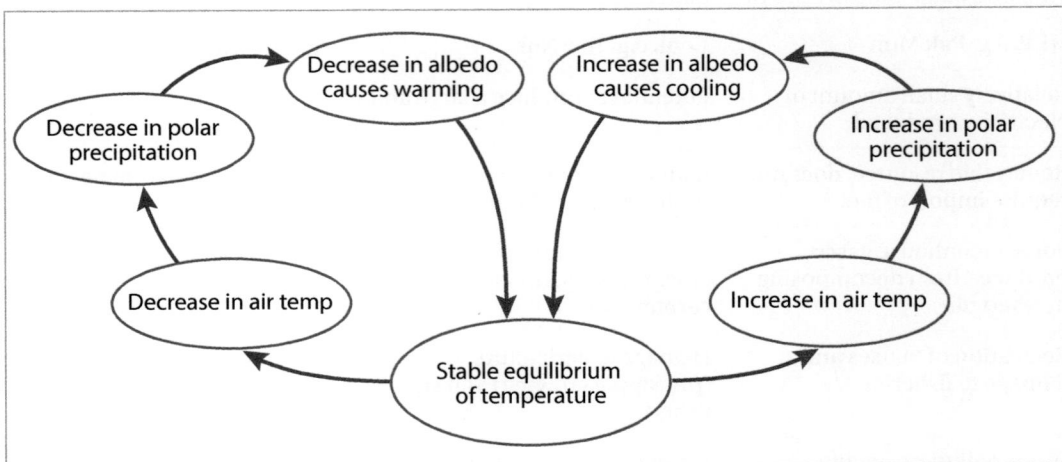

For an example of a positive feedback diagram, refer back Figure 1.6 on page 15.

8. Climate change predictions are difficult as they involve predictions of behaviour in complex systems. Interactions of the system components are often through feedbacks that have the ability to maintain or shift the equilibrium depending on many different variables and tipping points. Negative feedbacks may help regulate change, where as positive feedbacks lead to a more rapid effect. Climate change predictions also involve predicting the behaviour of people and societies in the future, including modelling demographic change, development and the outcomes of political and economic decision making. The quote also illustrates that scientists, aware of the uncertainties in computer models, still use them and predict continued warming of the globe: it is the best tool for the job.

9. See Table 7.8 on page 170.

10. Carbon dioxide is a naturally occurring gas that is returned through the atmosphere through respiration and combustion. Carbon is found in all living organisms, foods, and fuels. The release of this gas is, therefore, such a widespread and common place activity that the control of it is difficult. Carbon emissions are directly linked to the current models of economic development and controls are opposed by many vested interests as a result. Reducing emissions can be seen as a backward step in terms of economic development. Carbon-based economic development has been the model for society through history; developing alternatives has been slow. There is considerable cultural inertia regarding lifestyle change.

11. The Montreal Protocol (MP) has been seen by environmentalists as a success, the Kyoto Protocol (KP) a failure. The MP controls the production of chemicals that are entirely synthetic; the KP tried to control the release of naturally occurring emissions as well as synthetic. The MP did not need to control production in every country of the world as ODS manufacturing was only in some countries; the KP did not involve every country acting, yet all countries produce greenhouse gases.

12. See Table 7.9 on page 172.

13. Mitigation is the ultimate solution, effectively removing the source of the greenhouse gases. However, it has proven to be difficult to achieve in terms of the political will and the international cooperation required. Geoengineering may be seen as a more attractive option and does not require changes in lifestyle or policies, though it is a short-term measure: it buys time rather than

solves the problem. Geoengineering options can be cheaper, but may have negative side effects such as the potential to damage the ozone layer from artificial volcanic eruptions. Adaptation can be seen as living with, rather than doing anything about, climate change: the default position for a global society that does nothing about the change. The advantage of adaptation is that the solutions are tangible, with direct benefits that can be seen locally, such as coastal defence against rising sea levels.

14. Lifestyle changes are seen by many as an essential component of reducing emissions. However, it requires cooperation of significant amounts of people in order to have an effect. Many people reject lifestyle change, as they don't see it as effective or they don't believe that climate change is a problem in the first place. Traditional practices may be difficult to alter; cultural inertia may be a significant factor. The Kyoto Protocol demonstrated the difficulties of achieving a binding agreement that has significance. This agreement was seen as a step in the right direction, but was not agreed to by significant countries such as the USA. Even fully ratified the Kyoto Protocol would have a limited reduction of predicted warming in the future, tougher agreements would have even more resistance.

15. Comparisons of personal opinions will, of course, depend on your opinion, but you should make effective comparisons with other groups. For example, it could be a collection like this for someone who believes in ACC:

    (a) I agree with Gordon Brown's statement. The group that wishes to disagree does so in spite of evidence; their opinions seem to be due to political motivation and belief rather than scientific understanding.

    (b) I agree with Prof. Stott's statement, science does not work through consensus. However, it is the best understanding we have at the time, so politicians would do well to operate based on the current understanding of scientific consensus.

    (c) The surveys from the UK and the US show divided opinions, though the majority of people believe in ACC, including me. Political belief makes for considerable difficulties for some to accept the evidence of climate change, for example, politicians on the right are often seen to dispute the evidence, but they rarely accept the evidence provided. Accusations that there is a widespread conspiracy of socialist scientists have been made by sitting US presidents, including by Donald Trump and George W. Bush. Investigations have failed to find evidence of such political bias. Cultural background can make considerable difference to the acceptance of climatic change, e.g. a perspective of a deity having total control over fate. This belief can override science, in a similar way that creationist belief may be stronger than evolution. Science education can and should help to provide a basis in fact and evidence for environmental decision making. However, the quality and bias in education systems itself varies considerably. It is possible that some systems may simply present one side of the climate change debate, or try to avoid education about this at all.

Chapter 8

Human Population Dynamics

1. See Table 8.1 on page 180.

2. See table:

	% NIR
2010 – 2015	1.11
2015 – 2020	0.99
2020 – 2025	0.86
2025 – 2030	0.73
2030 – 2035	0.62
2035 – 2040	0.53
2040 – 2045	0.44

3. The projection involves predictions of future deaths, births, technology, medicine, life style/quality etc. There is a lack of certainty on this data in some countries, even historically or for the present. Predictions are always even more uncertain as they will involve unpredictable aspects of human behaviour and other events, e.g. medical developments.

4. Birth rate has proved to be the hardest to predict. It is particularly difficult to predict human reproductive behaviour, which was over-predicted by the UN for example during the 1990s where rapid urbanisation caused birth rate to fall much faster in developing countries than expected.

8.4.7 Resource Use in Society

1. See table:

	%NI	DT
1970 – 1975	1.94	36
1975 – 1980	1.77	40
1980 – 1985	1.76	40
1985 – 1990	1.75	40
1990 – 1995	1.54	45
1995 – 2000	1.36	51
2000 – 2005	1.26	56
2005 – 2010	1.18	59
2010 – 2015	1.11	63
2015 – 2020	0.99	71
2020 – 2025	0.86	81
2025 – 2030	0.73	96
2030 – 2035	0.62	113
2035 – 2040	0.53	132
2040 – 2045	0.44	159

2. 1960s (35.5 on average).
3. Oldest possible age for a man = 90 years. For a woman = 80 years.
4. 9 females.

ANSWERS TO REVISION QUESTIONS

5. Demographic transition is a model which refers to observed changes in populations and predicts a pattern of decline in mortality and fertility over time due to social and economic development. This model shows what happens to populations over time: the pattern has been shown to be generally true in many countries through their history and in the world as a whole. It is often shown as a figure: see Figure 8.3 on page 184.

6. Global doubling time is at its shortest when population is growing rapidly, as there is a large difference between birth rate and death rate. Though every country, and even populations within a country, may be in different stages of the DTM, countries and populations have slowly moved through the transition during the 20th century. Consequently, the model would suggest that many countries were in stage 2 when the population was growing most rapidly in the 1960s, 70s, and 80s. As the doubling time begins to increase and growth rate falls, it indicates more countries are moving out of stage 2 into stage 3.

7. 2.

8. Death rate falls first leading to rapid population growth.

9. 1.

10.

What are the limitations of knowledge?	The quality of data is very good in many countries, but some countries still have unreliable data. Predictions remain just that, and uncertainty remains high.
How reliable or representative is the data?	The model has been used to predict the behaviour of populations with some success. However it is based on empirical data, what has happened previously may not happen in the future as conditions change in unpredictable ways.
Is it an evidence based model that makes logical sense?	The model is based on considerable evidence, though the predictive elements–e.g. what happens in stage 5?– are unknown, there is no evidence for the future of course.
Are there feedbacks, with possible tipping points?	Yes there are feedbacks, e.g. urbanisation increasing, with a rural depopulation has produced a population with a lower fertility rate. A negative feedback loop. Feedbacks again are not always predictable and that gives a great deal of uncertainty.
Are there possible influences on the model that are missing?	Changes in medicine, education and quality of life in the future can all produce unexpected influence on future populations.
Is the model a useful tool for analysing the situation?	Yes, it remains a useful tool. But it is important to remember that the patterns in the DTM may not continue onwards into the future. Consequently the population predictions for 2050 vary by over 2 billion people.

11. Natural resources are naturally occurring goods and services. Goods are physical commodities and services are activities that are used as they are produced. Renewables are living resources produced by ecosystem productivity using solar energy. Non-renewable resources are those which decline in availability, permanently when used by a society, as the time scale of regeneration is geological.

12. Sustainability means living within the natural income without reducing the natural capital.

13. Any society, or segment of society, whose growth and development is dependent on reducing essential natural capital is not sustainable. However, this can be considered in relation to the dynamic nature of the resources; whilst societies may use resources up they may find alternative resources in the future.

14. See Table 8.3 on page 186.

15. This can be seen, by a historical example, when mineral oil saved the whale in the 1850s. Whales were being exploited to supply lamp oil, and many species came close to extinction. Developing the technological and scientific understanding to distil kerosene from petroleum and use it in

lamps reduced the demand for whale oil. During the time of the whale harvesting, the actions were unsustainable, but as they didn't continue in the long run they were sustainable. Other examples can be found, e.g. resource needs are changing rapidly in response to computing technology that requires, for example, rare earth metals that were not valued as a resource in the past. It is possible that new technological developments may mean they are not in demand in the future.

16. The intrinsic value of the whale would be our estimation of what the whale is worth, independent of how we value it as a resource. This is always going to be a difficult thing to establish, but even though it is hard to quantify it in economic terms, the value should be greater than zero. Most people accept that there is some kind of intrinsic value to existence for non-human species, as well as humans. The commercial value of the flesh is not related to this value, as it is determined by market interactions in supply, demand, and legal control. Consequently, it would be hard to compare the two values.

17. 20 whales.

18. 25 tonnes/ha/year.

Solid Domestic Waste/Human Population Carrying Capacity

1. In pre-Second World War USA during the Love Canal incident (8.3.5). Cultural factors: society seems to have taken an un-enquiring 'out of sight, out of mind' approach, where it was easy for companies to dispose of pollution in a hazardous way. Economic factors: considerable money was saved by using an unsafe disposal environment. Political factors reflected an unregulated approach to use of the environment.

2. Paper, glass, metals, plastics, organic waste, hazardous materials, e.g. electronic items, batteries, medicines, pesticides, cleaning products, etc. (not faeces).

3. See Table 8.4 on page 192.

4. Recycle: many materials can be recycled and, by closing a loop, it provides a resource, moving towards a circular economy. Organic recycling can lead to useful compost or methane digesters can be used to produce energy. Waste separation can help the recycling process, particularly in developed countries. In developing countries, waste may be recycled at a dump site by people who earn a living scavenging on the waste. This is a hazardous occupation, which may result in disease or injury. Arguably, it is therefore better to separate first. Incineration produces toxic air pollution and toxic cinder ash. Landfills may produce toxic leachate unless sealed, which is better.

5. Carrying capacity is defined as the maximum number of people that can be sustainably supported by a given environment. This figure is elusive to even experts as the total number includes a prediction of future behaviour, technological developments, and the dynamic resource demands by populations.

6. Carrying capacity can be increased by using new technology to increase productivity, e.g. through sustainable agricultural techniques that mean agriculture is not dependent on finite resources. Methods to reduce consumption, e.g. energy conservation techniques that reduce wasted energy from houses. Better government and fairer markets, which lead to a more equitable distribution of resources, e.g. reducing malnutrition caused by food distribution.

7. The ecological footprint is the area of land and water in the same vicinity as the population, which is required to absorb the population's pollution and waste and supply its resources at a given standard of living. Ecological footprints provide an ecological snapshot of current resource use and natural regeneration. Unlike carrying capacity these figures are more certain as they are not based upon any predictions.

8. (a) B; (b) A; (c) C.

9. Progress in sustainable development goals leads directly to reduction in death rates, some indirectly with the same effect. They also influence birth rates. Improving education can lead

to reduced birth rates, through raising awareness of choices of birth control and providing opportunities that prolong life in education, so reducing the number of marriages and families starting at a young age. This is particularly true when the education is of women and helps to empower women in society. Unplanned development can also influence population dynamics. Rapid urbanisation in LEDCs led eventually to a lowering of fertility rates and to large adjustments on projections in the 1990s, combined with death rate increases due to HIV/AIDS.

10. Cultural values are a powerful force in determining population growth. One reason why birth rate decline lags behind death rate decline in the DTM is that cultural attitudes to fertility are slow to change. Changes in death rate are of course easily accepted. Some arguments suggest that religious beliefs, acceptance of materialism and other fundamental values are the most significant in long term population growth.

11.

Brazil – 0.26 ha	Canada – 4.45 ha
More subsistence and extensive agriculture	Intensive agriculture
Lower per capita income and material consumption	Greater economic wealth and high consuming life styles
Lower carbon emissions per capita	Higher carbon emissions per capita
Carbon dioxide absorbed faster by tropical biome types	Lower carbon fixation tundra and grassland ecosystems

References

Adams, W. 2006. *The Future of Sustainability: Re-thinking Environment and Development in the Twenty-first Century*. Retrieved October 2021, from IUCN: https://www.iucn.org/downloads/iucn_future_of_sustanability.pdf

AmyBLUF. (6 December 2009). Gordon Brown Calls GW Skeptics 'Flat Earthers, Anti-Science'; Philip Stott Responds. Retrieved 29 December 2010, from YouTube: http://www.youtube.com/watch?v=iPpkMe8Z_Qc

Anderegg, W., Harold, J., Scneider, S., & Prall, J. (9 April 2010). 'Expert credibility in climate change'. *Proceedings of the National Academy of Sciences*, from https://www.pnas.org/content/107/27/12107

Anderson, E. (1993). *Plants and People of the Golden Triangle*. Portland: Dioscorides Press.
AQUASTAT. (November 2016). Water Withdrawal by Sector. Retrieved 17 May 2017, from AQUASTAT: https://www.fao.org/aquastat/en/

Beckwith, Paul (2019). 'Rise of The METHANE; Latest Science on Temporal and Spatial Variations' https://paulbeckwith.net/2019/03/01/rise-of-the-methane-over-time-and-latitude-two-videos/ (Accessed December 2021)

Begon, M., Harper, J., & Townsend, C. 2006. *Ecology: From Individuals to Ecosystems*. Oxford, UK: Blackwell.

Bonhommeau, S., Dubroca, L., Pape, O. L., Barde, J., Kaplan, D. M., Chassot, E., et al. (2013). 'Eating up the world's food web and the human trophic level'. *Proceedings of the National Academy of Sciences*, 20617–20620, from https://pubmed.ncbi.nlm.nih.gov/24297882/

Bookchin, M. (1989). 'Death of a small planet'. The Progressive, 19–23

BP (2017). *BP Statistical Review of World Energy*. London: BP, from https://www.connaissancedesenergies.org/sites/default/files/pdf-actualites/bp-statistical-review-of-world-energy-2017-full-report.pdf

Callicot, J. B. (1993). 'American conservation philosophy: a brief history'. ECOS 14 (1) pp. 41–46.

Canadell, P., & Dhakal, S. (2010). Global Carbon Project. Retrieved 19 December 2010, from Global Carbon Project: http://www.globalcarbonproject.org

Carson, R. (1963). *Silent Spring*. Boston: Houghton Mifflin.

Centre for Health, Environment and Justice. (n.d.). Love Canal fact pact. Retrieved November 2010, from Centre for Health, Environment and Justice, link updated October 2021: http://chej.org/wp-content/uploads/Love-Canal-Factpack-PUB-001.pdf

Cockburn, Harry (2021). 'The sea on fire, record-breaking floods and a heat dome: The biggest climate crisis moments of 2021'. *The Independent* 28th December 2021. https://www.independent.co.uk/climate-change/news/2021-heatwaves-floods-oil-spill-b1973110.html

(Accessed December 2021)

Cohen, J. (1995). *How Many People Can the Earth Support?* New York: W.W. Norton and Company.

Cohen, J. (2010). 'Solving the Resource-Population Equation in the Developing World'. Retrieved 28 December 2010, from the new security beat: http://newsecuritybeat.blogspot.com/2010/12/watch-joel-e-cohen-on-solving-resource.html

Connell, J. (24 March 1978). 'Diversity in Tropical Rain Forest and Coral Reefs'. *Science*, pp. 1302–1310.

Conservation International. (n.d.). Indo-Burma. Retrieved September 2010. www.conservation.org

Díaz-Pérez, Rodríguez-Zaragoza, Ortiz M, C.-M. A., JD, C., & Ríos-Jara. (31 August 2016). Biological, Ecological and Functional Diversity of Fish and Coral Assemblages in the Caribbean Sea. *PLoS*, from https://journals.plos.org/plosone/article?id=10.1371/journal.pone.0161812

EIA. (2008). *U.S. Energy Information Administration.* Retrieved 12 November 2010, from https://www.eia.gov/totalenergy/

Elliott, S. (2005). *How to Plant a Forest.* Chiang Mai: Forest Restoration Research Unit, Chiang Mai University.

Elton, C. (1958). *The Ecology of Invasions by Plants and Animals.* London: Methuen.

Erwin, T. (1988). 'The tropical forest canopy: The heart of biotic diversity'. In E. O. Wilson, *Biodiversity* (pp. 123–129). Washington: National Academy Press.

Etminan, M., G. Myhre, E. J. Highwood, and K. P. Shine (2016). 'Radiative forcing of carbon dioxide, methane, and nitrous oxide: A significant revision of the methane radiative forcing', *Geophys. Res. Lett.*, 43, 12,614–12,623 https://agupubs.onlinelibrary.wiley.com/doi/epdf/10.1002/2016GL071930 (Accessed December 2021)

Fahn, J. (2004). *A Land on Fire.* Chiang Mai, Thailand: Silkworm Books.

FAO. (2013). *Edible Insects: Future Prospects for Food and feed security.* Rome: Food and Agricultural Organisation of the United Nations.

Fluck, R. (1992). *Energy in Farm Production.* Atlanta: Elsevier Science.

Forsyth, T., & Walker, A. (2008). *Forest Guardians, Forest Destroyers: The Politics of Environmental Knowledge in Northern Thailand.* Seattle: University of Washington Press.

Forum for the Future. (2011). *What Is Sustainable Development?* Retrieved 12 April 2011, from Forum for the future: action for a sustainable world: https://www.forumforthefuture.org/

Fowles, J. (1992). *The Tree.* St. Albans: The Sumach Press.

Fox News. (29 May 2006). Retrieved 27 December 2010, from Think Progress: https://thinkprogress.org/

Fuller, R. Buckminster, (1969) *Operating Manual for Spaceship Earth.* Zurich: Lars Muller Publishers.

Global Footprint Network. (2011). *Data and Results.* Retrieved 2011, from Global Footprint Network advancing the science of sustainability: https://www.footprintnetwork.org/

Gore, A. (1992). 'Foreword: The Coming "Environment Decade"'. In A. Gore, *Earth in the Balance* (p. xxii). New York: Rodale.

Graham, M. (15 December 1998). 'Its Fate May Foreshadow Our Own'. *Bangkok Post.*

Henson, R. (2011). *The Rough Guide to Climate Change.* London: Rough Guides.

Huckle, J. (1996). 'Realising Sustainability in Changing Times'. In J. Huckle, & S. Sterling, *Education for Sustainability* (pp. 3–17). London: Earthscan Publications.

Hutchinson, R. E. (1957). Concluding Remarks. *Cold Spring Harbor. Symp. Quant. Biol. 22,* 415–427.

IEA. (2016). *Thailand Electricity Security Assessment.* Paris: International Energy Agency.

IPCC. (1990). *Climate Change: The IPCC Scientific Assessment.* Cambridge: Cambridge University Press.

IPCC. (2007). *Climate Change 2007: The Physical Science Basis.* New York: UN.

IPCC. (2013). *Climate Change 2013: The Physical Science Basis (AR5).* [Online] Available at: www.ipcc.ch

IPCC (2018). Summary for Policymakers. In Special Report: *Global Warming of 1.5°C.* [Masson-Delmotte, V., P. Zhai, H.-O. Pörtner, D. Roberts, J. Skea, P.R. Shukla, A. Pirani, W. Moufouma-Okia, C. Péan, R. Pidcock, S. Connors, J.B.R. Matthews, Y. Chen, X. Zhou, M.I. Gomis, E. Lonnoy, T. Maycock, M. Tignor, and T. Waterfield (eds.)]. World Meteorological Organization, Geneva, Switzerland. https://www.ipcc.ch/sr15/ (Accessed December 2021)

IPCC (2021). *AR6 Climate Change 2021: The Physical Science Basis.* Contribution of Working Group I to the Sixth Assessment Report of the Intergovernmental Panel on Climate Change [Masson-Delmotte, V., P. Zhai, A. Pirani, S.L. Connors, C. Péan, S. Berger, N. Caud, Y. Chen, L. Goldfarb, M.I. Gomis, M. Huang, K. Leitzell, E. Lonnoy, J.B.R. Matthews, T.K. Maycock, T. Waterfield, O. Yelekçi, R. Yu, and B. Zhou (eds.)]. Cambridge University Press. https://www.ipcc.ch/report/ar6/wg1/ (Accessed December 2021)

Kundstadter, P., Chapman, E. C., & Sabhasri, S. (1978). *Famers in the Forest: Economic Development and Marginal Agriculture in Northern Thailand.* Honolulu: University Press of Hawaii.

Kunzig, R. (2011). Population 7 Billion: How Your World Will Change. *National Geographic* January 2011, 32–69.

Leiserowitz, A., Maibach, E., & Roser-Renouf, C. (2010). *Climate Change in the American Mind: Americans' Global Warming Beliefs and Attitudes in January 2010.* New Haven: Yale University and George Mason Univerisity.

Lekagul, B., & McNeely, J. (1977). *Mammals of Thailand.* Bangkok: Saha Karn Bhaet Co.

Leopold, A. (1949). *A Sand County Almanac.* New York: Oxford University Press.

Liu, Y., Olaussen, J., & Skonhoft, A. (2010). 'Wild and Farmed Salmon in Norway – A Review'. *Marine Policy*, 1-6, http://folk.ntnu.no/skonhoft/Marine%20Policy%20salmon%20overview%202010.pdf

Lovelock, J. (1989). *The Ages of Gaia: A Biography of Our Living Earth.* Oxford, UK: Oxford University Press.

Lynas, M. (2008). *Six Degrees Our Future on a Hotter Planet.* London: Fourth Estate.

Lynas et al (2021). 'Greater than 99% consensus on human caused climate change in the peer-reviewed scientific literature'. *Environ. Res. Lett.* 16 114005.

MA. (2005). *Millennium Ecosystem Assessment.* Retrieved 15 July 2016, from http://www.millenniumassessment.org/en/index.html

Matsumoto, S., & Fukuda, K. (2002). *Thailand: Bor Nok Coal Plant FAQ.* Tokyo: Mekong Watch Japan.

Maxwell, J. (2004). A Synopsis of the Vegetation of Thailand. *The Natural History Journal of Chulalongkorn University*, 19–29, http://www.biology.sc.chula.ac.th/TNH/archives/v4_no2/4-2,19-29.pdf

May, R., & Lawton, J. (1995). *Extinction Rates.* Oxford: Oxford University Press.

Meadows, D. (2008). *Thinking in Systems.* Vermont: Sustainability Institute.

Metcalfe, L. (2010). *NationMaster.com.* Retrieved 22 October 2010, from http://www.nationmaster.com/index.php

Monbiot, G. (2007). *Heat: How We Can Stop the Planet Burning.* London: Penguin.

National Research Council. (2010). *Ecosystem Concepts for Sustainable Bivalve Mariculture.* Washington DC: The National Academies Press.

Naylor, R. G. R. et al. (2000). *Effect of Aquaculture on World Fish Supplies. Nature*, Volume 405, pp. 1017-1024.

Newby, H. (1987). *Country Life: A social history of rural England.* London: Cardinal.

Noble, R., Smith, P., & Pattullo, P. (2012). *Water Equity in Tourism.* London: Toruism Concern.

Odum, H. (1957). 'Trophic Structure and Productivity of Silver Springs, Florida'. *Ecol. Monogr.27* , 55-112.

ONS. (2015, July 13). *Sustainable Development Indicators: Office for National Statistics.* Retrieved 13 July 2016, from Office for National Statistics: https://www.ons.gov.uk/peoplepopulationandcommunity/wellbeing/articles/sustainabledevelopmentindicators/2015-07-13

Oreskes, N. (2004). 'Beyond the Ivory Tower: The Scientific Consensus on Climate Change'. *Science 306 (5702):1686,* https://science.sciencemag.org/content/306/5702/1686

O'Riordan, T. (1981). *Environmentalism.* 2nd edn. London: Pion.

OU. (1986). *Block A Ecosystems Unit 3 Decomposition and Mineral Cycling.* Milton Keynes, UK: Open University Press.

Page, M. L. (2011). 'Climate Change: A Special Report: What We Know...and What We Don't Know'. *New Scientist*, 22 October, pp. 36-43.

Pearce, D., Markandya, A., & Barbier, E. (1989). *Blueprint for a Green Economy.* London: Earthscan Publications.

Pimentel, D. (2009). 'Energy Inputs in Food Crop Production in Developing and Developed Nations'. *Energies*, Volume 2, pp. 1-24.

Ramage, M., & Shipp, K. (2009). *Systems Thinkers.* London: Springer.

Raup, D. (1988). 'Diversity Crises in the Geological Past'. In E. O. Wilson, *Biodiversity* (pp. 51-57). Washington: National Academy Press.

Reid, A. (1996, October). 'Exploring Values in Sustainable Development'. *Teaching Geography* , pp. 168-172.

Riley, C. (1996). 'Mammals and Other Animals'. *Field Studies* (8) , pp. 665-676.

Saipunkaew, W., Wolseley, P., Chimonides, P., & Boonpragob, K. (2007). 'Epiphytic macrolichens as indicators of environmental alteration in northern Thailand'. *Environmental Pollution* 146 , pp. 366-374.

Schneider, S. H. (1990). 'The Science of Climate-modelling and a Perspective on the Global Warming Debate'. In J. Leggett, *Global Warming: The Greenpeace Report* (pp. 44-67). Oxford: Oxford University Press.

Smith, R. (1995). *Ecology and Field Biology.* 5th edn. New York: Harper Collins.

Southward, A. (20 July 1978). 'Marine Life and the Amoco Cadiz'. *New Scientist*, pp. 174-176.

Spence, A., Venables, D., Pidgeon, N., Poortinga, W., & Demski, C. (2010). *Public Perceptions of Climate Change and Energy Futures in Britain: Summary findings of a survey conducted in January-March 2010.* Cardiff: Cardiff University: School of Psychology, https://doc.ukdataservice.ac.uk/doc/6581/mrdoc/pdf/6581final_report.pdf

Stern, S. N. (2006). *The Economics of Climate Change.* London: HM Treasury. https://personal.lse.ac.uk/sternn/108NHS.pdf

Thailand Development Research Institute. (2000). *The Pak Mun Dam, Mekong River Basin, Thailand.* Cape Town: World Commission on Dams.

Tollefson, J. (2010, July 1). 'An Erosion of Trust'. *Nature* Vol 466, pp. 24–26.

Tollefson, J. (2021). 'COVID curbed carbon emissions in 2020 — but not by much'. *Nature* Vol 589, 343

Tudge, C. (1991). *Global Ecology.* Verona: Arnoldo Mondadori.

Tudge, C. (2002). *Food for the Future.* London: Dorling Kindersley.

Tuffin, W., & Sae-Tang, N. (2008). *Ban Mae Lai Socio-economic Survey.* Chiang Mai: Track of the Tiger.

United Nations. (2016, March). *Sustainable Development Goal Indicators Website.* Retrieved 13 July 2016, from: http://unstats.un.org/sdgs/iaeg-sdgs/metadata-compilation/

United Nations. (2008). *World Population Prospects.* Retrieved 20 April 2011, from United Nations Department for Economic and Social Affairs: https://www.un.org/development/desa/pd/

United Nations (2021). Secretary-General's statement on the IPCC Working Group 1 Report on the Physical Science Basis of the Sixth Assessment. https://www.un.org/sg/en/content/secretary-generals-statement-the-ipcc-working-group-1-report-the-physical-science-basis-of-the-sixth-assessment (Accessed December 2021)

UNEP (2020). *Emissions Gap Report 2020.* Executive summary. United Nations Environment Programme, Nairobi. http://www.unenvironment.org/emissionsgap (Accessed December 2021)

Wilson, E. O. (1992). *The Diversity of Life.* London: Penguin.

World Commission on Environment and Development. (1987). *Our Common Future.* New York: Oxford University Press.

Worldwatch Institute. (2009). *State of the World: Into a Warming World.* New York: Norton and Company.

Worldwatch Institute. (2015). *Vital Signs Online.* Retrieved 13 July 2016, from Global trends that shape our future: http://vitalsigns.worldwatch.org/. The Worldwatch Institute was wound up in 2017, see https://www.commondreams.org/organization/worldwatch-institute.

Wright, R. T., & Nebel, B. J. (2002). *Environmental Science: Toward a sustainable future.* 8th edn. London, UK: Prentice Hall International.

Notes

Notes

Notes

Notes